I0038314

Frontiers in Biomaterials

(Volume 5)

(Stem Cell Biology and Regenerative Medicine)

Edited by

Mehdi Razavi

Department of Radiology, School of Medicine, Stanford University, Palo Alto, California 94304, USA

BENTHAM SCIENCE PUBLISHERS LTD.
End User License Agreement (for non-institutional, personal use)

This is an agreement between you and Bentham Science Publishers Ltd. Please read this License Agreement carefully before using the ebook/echapter/ejournal (**"Work"**). Your use of the Work constitutes your agreement to the terms and conditions set forth in this License Agreement. If you do not agree to these terms and conditions then you should not use the Work.

Bentham Science Publishers agrees to grant you a non-exclusive, non-transferable limited license to use the Work subject to and in accordance with the following terms and conditions. This License Agreement is for non-library, personal use only. For a library / institutional / multi user license in respect of the Work, please contact: permission@benthamscience.org.

Usage Rules:

1. All rights reserved: The Work is the subject of copyright and Bentham Science Publishers either owns the Work (and the copyright in it) or is licensed to distribute the Work. You shall not copy, reproduce, modify, remove, delete, augment, add to, publish, transmit, sell, resell, create derivative works from, or in any way exploit the Work or make the Work available for others to do any of the same, in any form or by any means, in whole or in part, in each case without the prior written permission of Bentham Science Publishers, unless stated otherwise in this License Agreement.
2. You may download a copy of the Work on one occasion to one personal computer (including tablet, laptop, desktop, or other such devices). You may make one back-up copy of the Work to avoid losing it. The following DRM (Digital Rights Management) policy may also be applicable to the Work at Bentham Science Publishers' election, acting in its sole discretion:

- 25 'copy' commands can be executed every 7 days in respect of the Work. The text selected for copying cannot extend to more than a single page. Each time a text 'copy' command is executed, irrespective of whether the text selection is made from within one page or from separate pages, it will be considered as a separate / individual 'copy' command.
- 25 pages only from the Work can be printed every 7 days.

3. The unauthorised use or distribution of copyrighted or other proprietary content is illegal and could subject you to liability for substantial money damages. You will be liable for any damage resulting from your misuse of the Work or any violation of this License Agreement, including any infringement by you of copyrights or proprietary rights.

Disclaimer:

Bentham Science Publishers does not guarantee that the information in the Work is error-free, or warrant that it will meet your requirements or that access to the Work will be uninterrupted or error-free. The Work is provided "as is" without warranty of any kind, either express or implied or statutory, including, without limitation, implied warranties of merchantability and fitness for a particular purpose. The entire risk as to the results and performance of the Work is assumed by you. No responsibility is assumed by Bentham Science Publishers, its staff, editors and/or authors for any injury and/or damage to persons or property as a matter of products liability, negligence or otherwise, or from any use or operation of any methods, products instruction, advertisements or ideas contained in the Work.

Limitation of Liability:

In no event will Bentham Science Publishers, its staff, editors and/or authors, be liable for any damages, including, without limitation, special, incidental and/or consequential damages and/or damages for lost data and/or profits arising out of (whether directly or indirectly) the use or inability to use the Work. The entire liability of Bentham Science Publishers shall be limited to the amount actually paid by you for the Work.

General:

1. Any dispute or claim arising out of or in connection with this License Agreement or the Work (including non-contractual disputes or claims) will be governed by and construed in accordance with the laws of the U.A.E. as applied in the Emirate of Dubai. Each party agrees that the courts of the Emirate of Dubai shall have exclusive jurisdiction to settle any dispute or claim arising out of or in connection with this License Agreement or the Work (including non-contractual disputes or claims).

2. Your rights under this License Agreement will automatically terminate without notice and without the need for a court order if at any point you breach any terms of this License Agreement. In no event will any delay or failure by Bentham Science Publishers in enforcing your compliance with this License Agreement constitute a waiver of any of its rights.

3. You acknowledge that you have read this License Agreement, and agree to be bound by its terms and conditions. To the extent that any other terms and conditions presented on any website of Bentham Science Publishers conflict with, or are inconsistent with, the terms and conditions set out in this License Agreement, you acknowledge that the terms and conditions set out in this License Agreement shall prevail.

Bentham Science Publishers Ltd.
Executive Suite Y - 2
PO Box 7917, Saif Zone
Sharjah, U.A.E.
Email: subscriptions@benthamscience.org

**BENTHAM
SCIENCE**

CONTENTS

FOREWORD

Biomaterials have come a long way since the first total joint replacements, which were introduced at a time when biomaterials were selected for their corrosion resistance. Orthopaedic surgeons initially selected materials which would stimulate the least reaction from the body. Materials used were "nearly inert" metal alloys and polymers. Total joint replacements revolutionised surgery and were life changing for patients. However, such materials are eventually rejected by the body, not in the same way as transplants, but because a thin layer of scar tissue forms around them, isolating them from the body, eventually causing the implant to be forced out of position. This became more problematic when clinicians attempted to repair or restore other parts of the skeleton or other tissues.

In 1969 (published in 1971), the invention of Bioglass® by Larry Hench, then at the University of Florida in Gainesville (USA), changed the face of orthopaedics. Bioglass was the first synthetic material that was found to bond with bone (no scar tissue). It is also biodegradable. However, it was not until the mid-1990s when the first Bioglass synthetic bone graft for bone regeneration reached the market. Now, it has been used in more than 1.5 million patients. Between the concept and clinical use of Bioglass, other bioactive ceramics reached clinicians first, such as synthetic hydroxyapatite, which is similar to bone mineral and also bonds with bone, albeit slower than Bioglass. This triggered the use of other calcium phosphate variants, such as tricalcium phosphate.

I mentioned that bioceramics can be biodegradable. This is possible by dissolution (also happens in water) or by cellular action (*e.g.* macrophages or osteoclasts). Hench termed the combination of biodegradation and bioactivity as 3rd Generation Biomaterials in a Science review in 2002.

Biomaterials are now being designed to deal with the body's own healing for many different clinical indications. To work well they must be used as temporary templates or scaffolding, specifically designed for the tissue that is being repaired. Scaffolds made of bioactive and biodegradable materials could present a 4th Generation if they are able to stimulate another course of action, *e.g.* blood vessel growth or bearing load. They can be labelled as 5th Generation, if they do both.

Remarkably, biomaterials have now gone beyond bone and orthopaedics. Almost every tissue in the body has received research attention, with clinical products at various stages of development. In this book, scaffolds for nerves, cardiovascular system, liver, kidney and skin applications are described in addition to bone, cartilage and dental. The translation of new devices from concept to clinic is a great challenge for biomaterials researchers, one that is certainly not lost on the authors of this book.

This book begins with important, and perhaps more conventional biodegradable materials, which are the biodegradable polymers that are used in sutures. Bioceramics are usually too brittle for load bearing structures that must take cyclic load, therefore, in this book they have been included within composites with polymers as the matrix. Metals are now also being made to be biodegradable.

Scaffolds are often designed to mimic the macrostructure of the host tissue, with blood vessels growing through the pore networks to feed the new tissue. Hydrogels are another important type of polymers which mimic the extracellular matrix of tissues. Hydrogels are particularly beneficial for cell types that exist in a 3D gel-like environment. Their unique property is their ability to transport nutrients through their watery networks to cells.

Scaffolds can be employed as an implant on their own, or can be seeded with cells (*e.g.* stem cells) *in vitro* prior to implantation, which is termed as tissue engineering.

The concept of bioactive, biodegradable and strong scaffolds is an important area in healthcare. The UK Government highlighted Eight Great Technologies in 2013, suggesting great need and opportunity for growth. Two of those are Advanced Materials and Regenerative Medicine. When new medical devices are created in the laboratory, they must be translated to clinic. In order to deliver these scaffolds, new manufacturing methods are also needed, such as Additive Manufacturing and 3D printing, which can create the required architectures and also promote reproducibility in large numbers.

Other aspects of technology transfer are the need to pass tests prescribed by regulatory bodies. The devices often highlight ambiguities in the tests, so new tests have to be developed. There is a large area of research in tests that can more closely assess the *in vivo* situation. While researchers often study how tissue specific cells respond to scaffolds, an important area often neglected is that how immune cells respond.

This new book provides a basic level of understanding of all of the above topics, starting from scaffold design; some key biomaterials; manufacturing techniques; to technology transfer aspects that include testing scaffolds both *in vivo* and *in vitro*. It provides the necessary foundation of science and technology. For the experienced researcher the book provides a comprehensive overview of the important current topics in the field. Happy reading.

Dr. Julian R. Jones
Department of Materials
Imperial College London
South Kensington Campus
London SW7 2AZ
UK
E-mail: julian.r.jones@imperial.ac.uk

PREFACE

Regenerative medicine an umbrella term given to varied approaches of replacing or repairing damaged or diseased organs offers a radical new method for the treatment of injury and disease. Regenerative medicine promises a more permanent solution other than the existing pharmaceutical products, and with the introduction of the first few achievements in the field, it progresses from the realms of science to the surgery. In future, we hope to further extend the scope from skin and bone to liver and heart; damaged organs are being made to regrow, or to be replaced with viable alternatives using stem cells. Although, our understanding of stem cell biology has increased rapidly over the last few years, the apparently tremendous therapeutic potential of stem cells has not yet been realized. To this end, many researchers continue to work in areas such as stem cell niche, reprogramming, nanotechnology, biomimetics and 3D bioprinting. Regenerative medicine is highly cross-disciplinary and serves as a bridge between the basic science to bioengineering and clinical medicine. The objective of this book was to capture and consolidate these research in identifying problems, offering solutions and providing ideas to excite further innovation in the stem cell biology and regenerative medicine field to help scientists, engineers and clinicians to design treatments for traumatic injury or degenerative diseases. This book covers the chapters by leading biologists, engineers and clinicians and therefore, has fundamental information that will be of use to all researchers dealing with the regenerative medicine strategies. In this book, recent advances in the basic knowledge of regenerative medicine involved in tissue damage and regeneration have been discussed with remarkable current progress in stem cell biology such that the vision of clinical tissue repair strategies is shown as a tangible reality. This is a reference book for undergraduate and graduate courses, bioengineers, medical students and clinical laboratories. Finally, the efforts of all the contributors and the publisher are appreciated.

Dr. Mehdi Razavi
Department of Radiology, School of Medicine
Stanford University, Palo Alto, California 94304
USA
E-mail: mrazavi2659@gmail.com

List of Contributors

Abdulmonem A. Alshihri
Department of Prosthetic and Biomaterial Sciences, College of Dentistry, King Saud University, Riyadh 11545, Saudi Arabia
Department of Restorative and Biomaterial Sciences, Harvard School of Dental Medicine, Boston 02115, USA

Ding Weng
Tissue Engineering Labs, VA Boston Healthcare System, Boston, USA
Department of Orthopedics, Brigham and Women's Hospital, Harvard Medical School, Boston, USA
Department of Mechanical Engineering, State Key Lab of Tribology, Tsinghua University, Haidian Qu, Beijing Shi, China

Fatemeh Khatami
Skin Research Center, Shahid Beheshti University of Medical Sciences, Tehran, Iran

Julio Aleman
Molecular and Cellular Biosciences, Wake Forest School of Medicine, Winston Salem, NC, USA
Wake Forest Institute for Regenerative Medicine, Winston-Salem 27101, NC, USA

Kai Zhu
Biomaterials Innovation Research Center, Division of Biomedical Engineering, Department of Medicine, Brigham and Women's Hospital, Harvard Medical School, Cambridge, MA, USA
Department of Cardiac Surgery, Zhongshan Hospital, Fudan University, Shanghai, China

Mahboubeh Nabavinia
Harvard-MIT Division of Health Sciences and Technology, Massachusetts Institute of Technology, Cambridge, USA
Biomaterials Innovation Research Center, Department of Medicine, Brigham and Women's Hospital, Harvard Medical School, Cambridge, USA
Chemical Engineering Department, Sahand University of Technology, Tabriz, Iran

Margaux Duchamp
Biomaterials Innovation Research Center, Division of Biomedical Engineering, Department of Medicine, Brigham and Women's Hospital, Harvard Medical School, Cambridge, MA, USA
Department of Bioengineering, École Polytechnique Fédérale de Lausanne, Lausanne 1015, Switzerland

Mehdi Razavi
Department of Radiology, School of Medicine, Stanford University, Palo Alto, California 94304, USA

Ming Yan
Department of Biomedical Engineering, College of life information science and instrument engineering, Hangzhou Dianzi University, Hangzhou 310018, China

Monireh Torabi-Rahvar
Department of Cancer Immunotherapy and Regenerative Medicine, Breast Cancer Research Center, IBCRC, Tehran, Iran
Liver and Pancreatobiliary Diseases Research Center, Digestive Disease Research Institute, Tehran University of Medical Sciences, Tehran, Iran

Ning Li State Key Laboratory of Medical Molecular Biology, Institute of Basic Medical Sciences, Chinese Academy of Medical Sciences, Beijing 100005, China

Perihan Selcan Gungor-Ozkerim Harvard-MIT Division of Health Sciences and Technology, Massachusetts Institute of Technology, Cambridge, MA, 02139, USA

Pingping Nie School of Life Sciences, Sichuan University, Chengdu 610005, China

Reza M. Robati Skin Research Center, Shahid Beheshti University of Medical Sciences, Tehran, Iran

Ruodan Xu Interdisciplinary Nanoscience Center (iNANO), Aarhus University, Gustav Wieds Vej 14, 8000 Aarhus C, Denmark

Runzhe Chen Department of Hematology and Oncology, Zhongda Hospital, School of Medicine, Southeast University, Nanjing 210009, China

Wanting Niu Tissue Engineering Labs, VA Boston Healthcare System, Boston 02130, USA Department of Orthopedics, Brigham and Women's Hospital, Harvard Medical School, Boston, MA, USA

Wenjin Shi School of Life Sciences, Sichuan University, Chengdu 610005, China

Yi-Nan Zhang Institute of Biomaterials and Biomedical Engineering, University of Toronto, 164 College Street, Toronto, ON, M5S 3G9, Canada

Yu Shrike Zhang Biomaterials Innovation Research Center, Division of Biomedical Engineering, Department of Medicine, Brigham and Women's Hospital, Harvard Medical School, Cambridge, MA, USA

Frontiers in Biomaterials, 2017, Vol. 5, 1-35

CHAPTER 1

Stem Cell-based Modalities: From Basic Biology to Integration and Regeneration

Ruodan Xu[1,§], Wenjin Shi[2,§], Pingping Nie[2,§], Runzhe Chen[3,§], Ning Li[4], Mehdi Razavi[5], Wanting Niu[6,7,*] and Abdulmonem Alshihri[8,9,*]

[1] *Interdisciplinary Nanoscience Center (iNANO), Aarhus University, Gustav Wieds Vej 14, 8000 Aarhus C, Denmark*

[2] *School of Life Sciences, Sichuan University, Chengdu 610005, China*

[3] *Department of Hematology and Oncology, Zhongda Hospital, School of Medicine, Southeast University, Nanjing 210009, China*

[4] *State Key Laboratory of Medical Molecular Biology, Institute of Basic Medical Sciences, Chinese Academy of Medical Sciences, Beijing 100005, China*

[5] *Department of Radiology, School of Medicine, Stanford University, Palo Alto, California 94304, USA*

[6] *Tissue Engineering Labs, VA Boston Healthcare System, Boston 02130, USA*

[7] *Department of Orthopedics, Brigham and Women's Hospital, Harvard Medical School, Boston 02115, USA*

[8] *Department of Prosthetic and Biomaterial Sciences, College of Dentistry, King Saud University, Riyadh 11545, Saudi Arabia*

[9] *Department of Restorative and Biomaterial Sciences, Harvard School of Dental Medicine, Boston 02115, USA*

Abstract: Stem cells have attracted great interest of biomedical scientists and clinicians due to their unique abilities of self-renewal and multipotential differentiation. With the most current technologies, stem cells have been isolated from almost all types of tissue, including embryonic stem cells, somatic stem cells, and induced pluripotent stem cells. The mechanisms of cells behavior have been fully studied. In combination with tissue engineering skills, stem cells have been investigated in a better environment by simulating the three-dimensional environment. However, the long-term safety and efficiency of stem cell-based outcomes should be further evaluated prior to any clinical application.

* **Corresponding author Abdulmonem A. Alshihri:** Restorative and Biomaterial Sciences, Harvard School of Dental Medicine, Boston, MA, 02115, USA, Department of Prosthetic and Biomaterial Sciences, College of Dentistry, King Saud University, Riyadh, 11545, Saudi Arabia; Tel: +966114677333; Fax: +966114679015; E-mail: monem.alshihri@post.harvard.edu
* **Co-corresponding author Wanting Niu:** Tissue Engineering Labs, VA Boston Healthcare System, Boston, USA; Department of Orthopedics, Brigham and Women's Hospital, Harvard Medical School, Boston, MA, 02115, USA; Tel: +1-617-637-6609; E-mail: wantingniubioe@gmail.com
§ These authors contributed equally to this chapter.

Mehdi Razavi (Ed.)
All rights reserved-© 2017 Bentham Science Publishers

Keywords: Dental stem cells, Embryonic stem cells, Induced pluripotent stem cells, Stem cells, Somatic stem cells.

INTRODUCTION

Stem cells are defined as undifferentiated cells that are capable of self-renewal and differentiating into various mature cells, to support an individual's postnatal life by replacing the aging cells and repairing injured tissue. When the organ injury is too severe for the body to recover, organ/tissue transplantation is the first considered strategy in current clinical practice. Due to the shortage of organ donors, tissue engineering strategies have been rapidly developed with translational and regenerative goals. One important objective of stem cell-based tissue engineering is to avoid immune rejection after transplantation.

In this chapter, we introduce the basic characteristics of stem cells commonly investigated in biomedical research. It provides a background for the other chapters on stem cell-based tissue engineering applications.

EMBRYONIC STEM CELLS

Embryonic stem cells (ESC) have three features, including infinite proliferation, self-renewal and pluripotency. The studies on ESCs started from teratomas and embryonal carcinoma (EC) cells in the 1950s and 1960s, respectively [1, 2]. In the 1970s, Kahan and Ephrussi established cell cultures from both testicular and embryo-derived teratocarcinomas [3]. In the 1990s, Thomson derived human embryonic stem cell lines from human blastocysts [4]. Currently, ESCs are widely applied to various disciplines, such as regenerative medicine or cell therapy [5 - 7], developmental biology and pharmacological applications [8]. Although ESC research is consistently the topic of ethical debates, ESCs have shown a vital use in different research and therapeutic modalities.

Pluripotency of Embryonic Stem Cells

ESCs can be derived from the inner cell mass of the blastocyst [4] and primordial germ cells [9, 10]. The cells from the human inner cell mass can differentiate into primary human embryonic lineages *in vitro* [11]. ESCs from a mouse model however may be maintained *in vitro* culture with leukemia inhibiting factor (LIF) and without feeder cells [12]. These *in vitro* differentiation models are usually designed to mimic early embryonic development. Therefore, the growth factors, extracellular matrix (ECM) components and signaling molecules are selected based on the knowledge of developmental biology. Recent research advances have revealed that the heparan sulfate is also involved in regulating ESC functions and the differentiation fate decision [13]. Many ECM-integrin interactions could also facilitate ESC differentiation [14].

ESCs can differentiate into various tissues originating from the ectoderm, mesoderm and endoderm. Regarding neuroectoderm lineages, ESCs can differentiate into midbrain neural cells, forebrain and midbrain tyrosine hydroxylase (TH)-positive neurons, neural crest, oligodendrocytes, motor neurons and keratinocytes [15]. For neural induction, different media and chemical inducers should be provided in various combinations and at different time points. Perrier *et al.* cultured the hESCs (lines H1, H9 and HES-3) on feeders and pretreated them with L-glutamin and β-mercaptoethanol for 16 days. Thereafter, signal sonic hedgehog (SHH), fibroblast growth factor (FGF)-8, brain-derived neurotrophic factor (BDNF), glial cell line-derived neurotrophic factor (GDNF), ascorbic acid, and dibutyryl cAMP were supplemented for 28 days, to form rosette structures. The cells were detached mechanically from the feeders and seeded in polyornithine/laminin-coated dishes in the presence of SHH, FGF-8, AA, and BDNF for one more week, followed by exposure to Ca^{2+}/Mg^{2+} free Hanks' balanced salt solution for 1 h and then cultured in the same medium for another week. Finally, ~30%-50% of all of the cells expressed TUJ1 (a neuron marker). Approximately 64%-79% of the TUJ1$^+$ cell population was also TH positive, 5% were serotonin-positive and 1%-2% were GABA-positive neurons. When cells experienced long-term (>70 days) culturing process, astrocytc-likc cells and O4 positive oligodendrocyte-like cells were detected in this system [16].

ESCs are capable of differentiating into mesoderm cells, such as cardiomyocytes, blood cells, skeletal muscle cells [17] and smooth muscle cells [18, 19]. Cerdan *et al.* reported that 5 ng/ml vascular endothelial growth factor (VEGF)-A$_{165}$ could selectively induce erythropoietic development from hESCs (H1 and H9 lines) in the presence of bone morphogenic protein (BMP)-4 and hematopoietic cytokines. These cytokines included stem cell factor (SCF), Fms-related tyrosine kinase 3 ligand (Flt-3L), interleukin (IL)-3, IL-6 and granulocyte-colony stimulating factor (G-CSF). VEGF-A$_{165}$ increased the co-expression frequency of CD34 and kinase insert domain receptor (KDR, a receptor of VEGF-A$_{165}$) after 15 days in culture. In addition, the expression of embryonic-globin genes were also upregulated, together with hematopoietic transcription factor SCL/Tal-1 [20]. With the purpose of inducing platelets from ESCs, Kawaguchi *et al.* first enforced overexpression of Gata2 on mouse ESCs by inserting a cDNA encoding Gata2 into the cells to obtain i*Gata*2-ESCs and used 1 μg/ml doxycycline (Dox) to induce transgene expression. Secondly, after 5 days of differentiation towards hemogenic endothelial cells (HECs) followed by a 7-day subculture on top of OP9 stromal feeder cells with 10 ng/ml mouse thrombopoietin (TPO) and Dox, the majority of the HECs robustly differentiated into megakaryocytes (Mks). Finally, 8 days after the initiation of the HEC culture, platelet-like cells were observed. These i*Gata*2-

ESC-induced platelets exhibited a similar morphology to peripheral blood platelets but were larger in size. They also expressed glycoprotein markers, *e.g.*, CD41 (GPIIb), CD42b (GPIb) and CD61 (GPIVa) [21]. As a natural substance found in grapes, resveratrol plays an important role in cardiovascular tissue protection. Ding *et al.* tested the feasibility of differentiating cardiomyocytes from ESCs by exposing mouse ESCs to different concentrations of resveratrol. Their results showed that 10 μmol/L was found to be safe and optimal to promote the mESC differentiation to cardiomyocytes [22].

For endodermal differentiation, ESCs could be induced to differentiate into hepatocytes [23] and pancreatic β-cells [24], which have potential applications for tissue regeneration. However, it is technically difficult to induce hepatocyte differentiation because the related molecular mechanisms have not been fully understood. Ishii *et al.* differentiated murine ESCs to endodermal cells or hepatic progenitor cells. And then, co-cultured these cells with MSCs derived from fetal liver mesenchymal cells to make the progenitor cells undergo further differentiation to mature hepatocytes through cell-to-cell contact. These resulting cells exhibited ammonia removal activity, albumin secretion ability, glycogen synthesis and storage, and cytochrome P450 enzymatic activity [23]. ESC-derived insulin-producing cells have been widely studied for diabetes treatment. Brolen *et al.* found that spontaneous differentiation of hESCs under 2-D growth conditions could result in Pdx1(+)/Foxa2(+) pancreatic progenitors. Cotransplantation of differentiated hESCs with mouse embryonic dorsal pancreas cells led to further differentiation of β-cell-like cells. These cells share many properties with normal β cells, including the synthesis of insulin and nuclear localization of key β-cell transcription factors: Foxa2, Pdx1, and Isl1 [24].

Regarding tissue engineering applications, the way in which physical cues, such as the stiffness of biomaterials, influences ESC differentiation has also been reported. Alginate hydrogels, with Young's moduli in the range of 242 to 1337 Pa, were employed for the investigation of murine ESC initial differentiation and gene expression profiles. The expression of mesodermal lineage markers varies in response to the stiffness changes of the gels. For example, FGF-8 had ~10-fold upregulation when using gels in the range of 650 to 950 Pa. In a lower range of 500 to 850 Pa, an endodermal marker, CXCR4, showed a 30 to 50-fold increase, and AFP exhibited a 90-fold increase in gene expression [25].

Establishment of Embryonic Stem Cell Lines

In 1998, Thomson derived human ESC (hESC) lines including H1, H7, H9, H13 and H14, which retain developmental potency for differentiating into trophoblast, endoderm, mesoderm and ectoderm cells [4]. In these cell lines, H1 and H9 have the normal karyotype of XY and XX, respectively and have been mostly used from 1999 to 2008 [26]. At present, more than 1,200 new human ESC lines have been created globally with various human leukocyte antigen (HLA) types and ethnic groups. A recent survey reported that the quality and developmental stage of embryos, isolation strategies of inner cell mass (ICM) and the culture media, are the four critical factors for hESC line establishment. The ideal conditions for hESC derivation have not been consolidated yet, and all of the lines display significant differences from the murine counterpart in epigenetic stability and morphology [27]. In spite of the enormous contribution of hESC research, the opportunities for U.S. scientists to study human ESCs were curtailed with the announcement from President George W. Bush that studies on cells started after August, 2001, would not be supported by federal grants. Due to the limitations on using federal funds to pursue genetic questions in human ESCs, the U.S. scientists can only employ the 21 lines listed on the NIH registry, which were developed with bovine serum and a limited genetic diversity [28].

New Strategies for Deriving Embryonic Stem Cells

Haploid cells are good materials for genetic analysis, whereas analysis is difficult to do with oocytes and sperm *in vitro*. To solve this problem, a series of studies have been conducted to establish haploid cell lines for genetic analysis. Modlinski [29], Tarkowski [30], Kaufman [31], Latham [32], and Yang [33] produced mouse haploid embryos, of which Yang and his co-workers [33] successfully established five mouse haploid embryonic stem cell (haESC) lines from androgenetic (AG) blastocysts by nuclear transfer techniques. The authors injected a haploid sperm head from the *Oct4-enhanced green fluorescent protein (EGFP)* transgenic mouse (C57BL/6 background) into an enucleated oocyte. Another set was created by removing the female pronucleus from oocytes and fertilized by *Actin-EGFP* transgenic male mice. After multiple rounds of FACS and passaging, the haploid cells were enriched. Finally, the derived AG-haESCs were expanded *in vitro* for 30 more passages, maintaining paternal imprints, expressing classical ESC pluripotency markers and differentiating into various types of tissues. In addition, injection of the haESCs into metaphase II (MII) oocytes can generate fertile mice. However, no AG-haESC lines with a Y-chromosome were observed in this study.

In humans, somatic cell nuclear transfer (SCNT) has been envisioned as an important approach for deriving patient-specific ESCs, which have the potential application for cell-based therapies. The largest barrier for deriving human NT-ESCs is the absence of activated critical embryonic genes from the somatic donor cell nucleus, so that the embryos fail to develop beyond the eight-cell stage [34, 35]. In 2013, Tachibana *et al.* [36] made an important breakthrough in this process. They reported that critical reprogramming factors in human MII oocytes are physically related to the chromosomes or spindle apparatus, which are depleted after enucleation. Therefore, they made a few improvements in the SCNT protocol, including the use of inactivated hemagglutinating virus of Japan (HVJ-E) to fuse nuclear donor cells with enucleated oocytes. It further stimulates the fused embryos with electroporation to activate them before exposing them to the standard ionomycin/DMAP (I/DMAP) activation. Approximately 10% of SCNT embryos could finally reach the blastocyst stage. Adding 10 nM trichostatin A (TSA) contributes to the achievement of stable NT-ESC lines, while adding 1.25 mM caffeine during spinal removal and fusion enhanced the blastocyst development rate and ESC line derivation. All of the derived cell lines expressed OCT-4, NANOG, SOX2, SSEA-4, TRA-1-60, and TRA-1-81. The efficiency of this method, by which NT-ESC lines can be derived from two oocytes, was very high. However, the donor cells are from fetal skin cells and skin fibroblasts from an 8-month-old patient with Leigh syndrome [37]. Therefore, it is necessary to further study if NT-ESC lines can be derived from adult somatic cells.

Embryonic Stem Cells *versus* Induced Pluripotent Stem Cells

Induced pluripotent stem cells (iPSCs) have similar gene expression profiles and developmental potential as ESCs. iPSCs were derived from somatic fibroblasts for the first time by enforcing expression of four transcription factors (Yamanaka factors), octamer 4 (Oct4), sex-determining region Y-box 2 (Sox2), Kruppel-like factor 4 (Klf4), and c-Myc [38]. Choi *et al.* recently reported that hiPSCs are similar to hESCs in their function by comparing genetically matched hESC with hiPSC lines [39]. ESCs have been applied to devastating and currently incurable diseases including spinal cord injury [5], Parkinson's disease [6], retinal degenerations [7], and type 1 diabetes [40]. In addition, ESCs also have considerable value for the development of biology research [41] and drug discovery [8]. Meanwhile, iPSCs can also be widely used in regenerative medicine, disease modeling, and drug discovery [42]. Therefore, the dilemma arises if iPSCs can replace ESCs in disease modeling and clinical applications in the future [38].

Importantly, the use and potential damage of human embryos for the derivation of human embryonic stem cells have evoked drastic ethical debates. iPSCs do not have this ethical issue as they are reprogramed from somatic cells. Moreover, it is more difficult to study ESCs compared to iPSCs. It is that the extraction and derivation of ESCs are limited, while the derivation of iPSCs from somatic cells is relatively less complicated [37].

It is dangerous to have undifferentiated stem cells, such as ESCs and iPSCs, under differentiated derivatives for transplantation because, with gene manipulation, the cells may form a teratoma [43]. The reprogramming process may cause genomic instability and abnormalities. Therefore, NT-ESCs and iPSCs may be more likely to cause tumorigenesis. In addition, mutations that are detected in human iPSCs do not necessarily exist in human ESCs [44 - 47].

As a result of the differences in the genotype between blastocyst-derived hESCs and the cells of a patient, hESCs commonly cause an immunological rejection. Whereas iPSCs are patient-specific and do not lead to an immunological rejection [36].

SOMATIC STEM CELLS

Somatic stem cells (SSCs) are self-renewable, multipotent cells with the ability to differentiate into several restricted lineages. They can be found in a variety of children and adult tissues, and have been known as adult stem cells [48]. The role of SSCs is to maintain and repair injured tissue. Typically, SSCs are named on the basis of the organ from which they are derived (such as haematopoietic stem cells) [49]. Here, we have outlined a few sources that have been considered for tissue engineering and therapeutic uses.

Mesenchymal Stem Cells

Mesenchymal stem cells (MSCs) are prototypical multipotent adult stem cells that were initially isolated and characterized by Friedenstein and his colleagues in 1970. Their observation was based on the tight adherence of cells to tissue culture surfaces and formation of fibroblast colonies [50]. MSCs were originally found in bone marrow (BMSCs). However, they have been isolated from many other tissues in humans such as adipose tissue (AMSCs) [51], cartilage [52], peripheral blood [53], umbilical cord [54], placenta [55], and synovial tissue [56]. MSCs isolated from various adult tissues express different morphology, differentiation potential, and gene expression profiles in standard culture conditions [57, 58]. It is common to classify MSCs based on their origin tissue, such as BMSCs for bone marrow-derived MSCs and AMSCs for adipose-derived mesenchymal stem cells [53, 59]. Following *in vitro* isolation and expansion, MSCs have been defined by their expression of various surface markers including CD105, CD73 and CD90 and the lack of expression of CD45, CD34, CD14 or CD11b, CD79a or CD19, and HLA-DR [58, 60]. Consequently, MSCs are isolated and highly enriched by their cell surface markers using immunostaining and cell sorting technologies. In terms of stemness, MSCs possess the ability to differentiate into multiple cell types that are specific for different tissues, including adipocytes, chondrocytes, osteocytes, and myocytes [59]. These specialized cells have their own characteristic morphologies, structures and functions, and each belongs to a particular tissue.

Compared to ESCs and iPSCs, MSCs are free of ethical concerns and have a low risk of forming teratomas and other types of tumors, as well as low immunogenicity [61]. In addition to their multilineage differentiation potential, MSCs have been widely used for cell-based tissue repair and tissue engineering. When MSCs are transplanted directly, as many as 90% of them would die in short time due to the inappropriate microenvironment, such as physical stress, inflammation, and hypoxia. Therefore, the cells could not be efficiently delivered and exert their functions on the damaged tissues. Advanced techniques for cell delivery are required, whereby MSCs can be incorporated into three-dimensional scaffolds that mimic the microenvironment in the tissue of the body. These scaffolds retain the cells, improve cell survival and assist with the integration into the host tissue. The results obtained from animal models have shown that MSCs are promising in the treatment of numerous diseases, mainly tissue injury and immune disorders. Clinical studies using MSC treatment are still in their infancy, and more work is needed before such therapies can be used routinely in patients. Several possibilities for their use in the clinic are currently being explored for safe and effective new treatments in the future [62 - 64].

Neural Stem Cells

Neural stem cells (NSCs) are stem cells derived from the central nervous system (CNS) that can self-renew and give rise to differentiated progenitor cells through asymmetric cell division to generate lineages of neurons as well as glia cells, such as astrocytes and oligodendrocytes [65]. In 1961, the first evidence of neurogenesis was reported in the adult mammalian brain using [3H]-thymidine incorporated into the DNA of dividing cells to study proliferation [61]. Neurogenesis from endogenous NSCs was primarily identified to occur mainly in two regions of the adult brain, the subgranular zone (SGZ) and the subventricular zone (SVZ) of the dentate gyrus (DG) in the hippocampus and olfactory bulb (OB), respectively. Other areas of the adult brain exhibit low levels of neurogenesis [66, 67].

In the SVZ, NSCs were shown to be genetically predetermined to generate specific subclasses of olfactory interneurons [68]. The mechanism of neurogenesis in normal conditions is not fully understood and still under intense investigation. The molecular control of fate determination from postnatal NSCs shares many aspects with fate determination in embryonic development. Because adult NSCs are normally found in a quiescent state, regulatory pathways can affect adult neurogenesis in ways that have no clear counterpart during embryogenesis. Bone morphogenic protein (BMP) signaling, for instance, regulates NSC behavior both during embryonic and adult neurogenesis. However, this pathway maintains stem cell proliferation in the embryo, while it promotes quiescence to prevent stem cell exhaustion in the adult brain [69]. Two major signaling pathways, Notch and Wnt, are involved in the regulation of NSC quiescence [70]. It is still unclear which molecules could activate NSCs. Generally, NSCs are activated in pathological conditions such as neurodegenerative diseases [71] or brain injury [72], but they are not altered in the same way.

NSCs isolated from OB (OBNSCs) could be used for autotransplantation due to their biosafety and histocompatibility properties. It was reported that the OBNSCs are less likely to form tumors compared to other stem cells when they were transplanted into CNS on animal models. It is relatively easier to obtain these cells through the nasal cavity via minimally invasive surgery. Therefore, OBNSCs are considered a good cell resource for the treatment of neurodegenerative conditions. Compared to SVZ-NSCs, OBNSCs demonstrated similar positive percentage of cells which express stem cell markers, *e.g.* nestin, SOX2, CD133 [73]. When compared to ESCs, the expressions of epigenetic-related transcription factor genes are highly increased in OBNSCs. However, the expressions of TUJ1, glial fibrillary acidic protein (GFAP), microtubule-associated protein (MAP) and O4 showed a lower positive ratio [74]. After the transplantation of OBNSCs into rat spinal cord injury (SCI) model, the cells differentiated into oligodendrocytes, astrocytes and neurons, and integrated into both grey and white matter with normal morphology. However, this study showed negative results of behavior tests in restoring the lost sensory and motor functions [75].

NSCs can give rise to different functional neuronal sub-types that use a variety of neurotransmitters such as dopamine, γ-aminobutyric acid (GABA), glutamate, and nitric oxide (NO) [76]. In addition, growth factors and other extrinsic signals could cause or induce neurons of different types. For example, epidermal growth factor (EGF), fibroblast growth factor 2 (FGF2) and brain-derived neurotrophic factor (BDNF) [77]. Studies on the genetic control of NSC-fate commitment may also have important therapeutic implications. NSCs' fate could be programed *in vivo* and the NSCs' progenies could be used for cell replacement in response to a lesion or disease. Finally, these studies may also be used to refine the *in vitro* protocols for the differentiation of NSCs into desired neuronal phenotypes and provide a promising approach for neural regeneration.

Satellite Cells

Satellite cells, or myosatellite cells, are quiescent muscle precursor cells in adult muscles. They were discovered through two independent studies in 1961 by Katz and Mauro [78, 79]. They are normally quiescent in adult muscle, located between the sarcolemma and basement membrane of terminally differentiated muscle fibers [80]. As a reserve population of cells, satellite cells have the potential to provide additional myonuclei to grow myoblasts in response to muscle injury. These cells can also give rise to regenerative muscle cells or return to a quiescent state to proliferate more satellite cells [81]. The recent discovery of a number of markers expressed by satellite cells has provided evidence that satellite cells differ from muscle stem cells [82]. It is possible that a sub-population of satellite cells may be derived from a more primitive stem cell. Satellite cell-derived muscle precursor cells may be used to repair and regenerate damaged or myopathic skeletal muscle or to act as vectors for gene therapy. In adults, satellite cells can be recruited to supply myoblasts for routine muscle fiber homeostasis or for the more sporadic demands of myofibre hypertrophy or repair [81]. The idea that satellite cells' function as myogenic precursors was initially based on studies of the distribution of labeled thymidine in growing or regenerating muscles. It led to the commonly accepted view that satellite cells divide to provide myonuclei for myofiber growth before becoming mitotically quiescent in normal mature muscles. Recently, transplantation of such single myofibers into muscles has provided good evidence that the satellite cells indeed act as myogenic stem cells *in vivo*. They are also able to give rise to new myofibers and, importantly, are able to renew themselves [83]. The satellite cell therefore fulfills the basic definition of a stem cell in that it can give rise to a differentiated cell type and maintain self-renewal process [84].

Cardiac Stem Cells

A population of resident cardiac stem cells (CSCs), known as cardiogenic progenitor cells (CPCs), has been found intimately connected by gap junctions to myocytes and fibroblasts as the supporting cells within the cardiac niches in the heart [85 - 87]. They are self-renewing, clonogenic, and multipotent. As well as giving rise to three major cardiac cell types: cardiomyocytes, vascular smooth muscle, and endothelial cells *in vitro* and *in vivo* [85]. These CSCs are thought to account for the physiological turnover of cardiac myocytes and vascular endothelial cells, which occurs in the heart in the absence of injury [88]. CSCs comprise less than 1% of the cells in the heart and have been subclassified according to their expression of surface marker transcription factors. These subclassifications are, c-kit-positive cells, cardiosphere-derived cells, stem cell antigen (Sca)-1-positive cells, islet 1-positive cells, stage-specific embryonic antigen (SSEA) 1-positive cells and Wt1 epicardial-derived cells [89 - 93]. Thus, the markers that show the phenotype of the CSC population are varied among different researchers. The existence of each cell population opens up additional opportunities for cardiac repair, especially in patients with ischemic cardiomyopathies.

Stimulation of endogenous regenerative activity is an attractive strategy but is limited to myocardial regeneration and repair. Drugs, growth factors or cytokines have been used to stimulate endogenous cardiomyocytes or CSCs *in situ*. For example, local delivery of biotinylated insulin-like growth factor (IGF)-1 complex increased cardiomyocyte growth both *in vitro* and *in vivo* [94, 95]. Alternatively, regeneration can be achieved by *ex vivo* propagation of CSCs followed by transplantation of the cells into the injured area. The anatomical source of the CSCs in the heart is an important factor for understanding their role in maintaining cardiac homeostasis and therapeutic potential. A proteolysis enzyme cocktail can digest small biopsies of the myocardium. It is followed by proliferation and expansion of CSCs in culture, and then these cells are transplanted into the injured area in the end. Animal studies have demonstrated that transplantation of c-kit-positive cells into the heart could enhance the formation of new blood vessels and myocardium [85]. MSCs are a promising source for cell-based treatment of myocardial infarction (MI). However, they contribute to scar formation without enhancing cardiac function. Thus, another approach for CSC activation was done by co-culturing CSCs together with MSCs [96]. To become a feasible option for clinical trials, cardiac cell therapy should adopt improved methods of cell survival and engraftment after transplantation. It can be done by the addition of growth factors [97, 98], genetic engineering and restoration of the damaged area with CSC-loaded scaffold constructs [99].

Liver Progenitor Cells

Liver stem/progenitor cells (LPCs) are often referred to as oval cells in rodents, which were first described by Farber using a rat model of liver carcinogenesis [100]. They are thought to reside within the terminal bile ductules (Hering Canals) located at the interface between parenchyma and biliary tracts. LPCs are activated and differentiated to hepatocytes and cholangiocytes, leading to functional recovery of the organ upon severe or chronic liver injury. Many groups have succeeded in isolating LPCs from the adult liver based on marker gene expression and flow cytometric cell purification followed by *in vitro* cultivation. Potential LPCs are usually identified as those that are positive for cholangiocyte markers with the expression of EpCAM, CD133, MIC1-1C3, and CK19 in mice, rats and humans [101].

Numerous molecular factors contribute to LPC activation or expansion either by direct or indirect signals. Among those factors, FGF7 and tumor necrosis factor (TNF)-related weak inducer of apoptosis (TWEAK) are of a particular interest for their capability to induce *de novo* activation of LPCs [102]. Other growth factors, such as hepatocyte growth factor (HGF) and EGF, have also been implicated in regulating proliferation and/or differentiation of LPCs [102 - 105]. These factors are mainly secreted from the surrounding environment and other cell types that interact with LPCs. Stellate cells have been suggested to physically interact with LPCs, provide HGF and promote pericellular collagen deposition to support LPC expansion [106]. These factors could be applied to cells in potential cell-based therapies to encounter liver disease by enhancing the regenerative capacity within the organ. However, patient with end-stage liver failure often require liver transplantation, for which donors are of a short availability with the concomitant risks of rejection. Similar with other types of adult stem cells, the very few resident endogenous LPCs and poor donor engraftment and survival are limited to the repair and restoration of the function of the injured liver. Development of LPC-based liver support constructs provides a promising alternative to liver transplantation for patients, especially those with acute hepatic failure (AHF) and end-stage chronic liver disease [107].

Dental Stem Cells

With the discovery of stem cells in teeth, several populations of cells with stem cell properties have been isolated from different parts of the tooth. All of these cells share a common lineage of being derived from neural crest cells and have generic mesenchymal stem cell-like properties. At least five types of dental MSC-like cells have been reported to differentiate into odontoblast-like cells. Those include dental pulp stem cells (DPSCs), stem cells from exfoliated deciduous teeth (SHED), periodontal ligament stem cells (PDLSCs), stem cells from apical papilla (SCAP), and dental follicle progenitor cells (DFPCs) [108]. The cells residing in different parts of the tooth play different roles in the formation of relevant tissues *in vivo*. Similarly, they have been shown to have different growth rates, marker gene expression and differentiation potential *in vitro*.

Various growth factors have been shown to regulate dental stem cells, such as FGF2, vascular endothelial growth factor (VEGF), or platelet-derived growth factor (PDGF), and bone morphogenetic protein-7 (BMP7). Such growth factors play a major role in neovascularization, neodentin or newly dental pulp tissue formation [109]. Recently, dental stem cells have been considered to be potential sources of cells for tissue regeneration and engineering. Studies have shown its positive outcomes in areas of repairing damaged tooth tissue, such as dentine, periodontal ligament and dental pulp, as well as other tissues [110].

INDUCED PLURIPOTENT STEM CELLS

Embryonic stem cells have the ability to differentiate into all cell types and proliferate rapidly after maintaining pluripotency. Human ESCs have been considered promising sources in cell transplantation therapies for various diseases and injuries. For example, spinal cord injury, myocardial infarction, type I diabetes, and muscular dystrophy. However, the clinical application of human ESCs faces ethical issues, as well as possible tissue rejection following implantation. One way to solve these problems is to generate pluripotent stem cells directly from somatic cells.

In 1960, Gurdon *et al.* generated tadpoles by transferring the nuclei of intestinal cells from an adult frog into oocytes [111]. His successful cloning showed that pluripotency-inducing factors do exist. In 2006, iPSCs were generated from adult fibroblasts by the retrovirus-mediated introduction of four transcription factors including Oct3/4, Sox2, c-Myc, and Klf4 [112]. These iPSCs are similar to ESCs in morphology, proliferation, and teratoma formation. In addition to fibroblasts, iPSCs have been generated from human hair follicle mesenchymal stem cells [113], nasal epithelial cells [114], neural stem cells [115], and hepatocytes [116]. These data demonstrated that pluripotency could be induced in various somatic cells, using only a few defined factors. A recent progress involving the derivation of iPSCs from malignant cells has opened a new door for cancer investigation [117].

Induction of Pluripotent Stem Cells from Mouse Embryonic and Adult Fibroblasts

In 2006, Takahashi *et al.* generated iPSCs successfully from mouse embryonic or adult fibroblasts by retrovirus-mediated introduction of four transcription factors. Those factors are including Oct3/4, Sox2, c-Myc, and Klf4 [112]. They selected 24 genes as candidates for factors that induce pluripotency in somatic cells. It was based on their hypothesis that such factors play pivotal roles in the maintenance of ESC identity.

Fbx15, as an undifferentiated ESC marker, is specifically expressed on ESCs and early embryos. However, it is dispensable for the maintenance of pluripotency and mouse development. It was expected that even partial activation of the Fbx15 locus would result in resistance to normal concentrations of G418. They therefore developed an assay system where the induction of the pluripotent state could be detected as resistance to G418 to evaluate these 24 candidate genes. At first, they introduced each of the 24 candidate genes into mouse embryonic fibroblasts (MEFs) from Fbx15bgeo/bgeo embryos by retroviral transduction. However, they did not obtain drug-resistant colonies with any single factor, indicating that no single candidate gene was sufficient to activate the Fbx15 locus. In contrast, transduction of all 24 candidates together generated G418-resistant colonies.

Secondly, to determine which genes of the 24 candidates are necessary for reprogramming, they examined the effects of the withdrawal of individual factors from the pool of transduced candidate genes on the formation of G418-resistant colonies. Finally, this research showed that Oct3/4, Klf4, Sox2, and c-Myc play important roles in the generation of iPSCs from MEFs. After subcutaneous injection of iPS-MEF4 into nude mice, histological examination revealed that the clone differentiated into all three germ layers, including neural tissue, cartilage, and columnar epithelium. This demonstrates that the majority of iPS-MEF4 clones exhibit pluripotency [112].

Improving the Reprogramming Rate of Induced Pluripotent Stem Cells

Although iPSCs have made prominent progress in various aspects, a significant limitation with using iPSCs in clinical treatment is the low translation efficiency of the cells. To improve the translation efficiency of the iPSCs, mainstream research on stem cells is focused on regulatory factors in the cell nucleus and high-throughput screening of small molecules. Researchers from Kyoto University found that the translation rate of iPSCs was significantly improved by reducing the oxygen concentration of the culture, and proved that the volume fraction of 5% oxygen concentration is the most appropriate [118]. It was also found that adding an antioxidant, such as Vitamin C, into the culture medium could improve the efficiency of somatic cell reprogramming [119]. Inhibiting the expression of p53 genes also helps to improve the reprogramming rate, which has been tested on mouse embryonic fibroblasts and colon cancer cells HCT116 [120].

Mesenchymal Stem Cells Derived from Induced Pluripotent Stem Cells

The use of mesenchymal or stromal stem cells in cancer patients or cancer survivors is a promising strategy to improve treatment of advanced cancer [121] and to repair tissue damage from cancers or radical therapies [122]. In general, it is easy to isolate the organization source of MSCs from the body and to expand them *in vitro*. However, bone marrow derived MSCs (BMSCs) and adipose derived stem cells (ASCs) have had many concerns in preclinical and animal research. The proliferation ability of MSCs is limited *in vitro*, which makes it challenging to meet the demand for high cell numbers for clinical applications. In addition to MSCs' limited proliferation potentials, they can lose some of their characteristics and functions. These changes could involve cellular phenotype, genotype, differentiation, cell cycle, intracellular reactive oxygen species (ROS) levels or any other traits [123, 124]. Moreover, tissue-derived MSCs may promote cancer progression and have considerable donor variations [125].

On the other hand, iPSCs proliferate with impunity in theory, meeting the demand for the large number of cells. However, the self-renewal and pluripotency of iPSCs may also lead to tumorigenesis and instability after transplantation *in vivo*. Sánchez [126] reported that inhibition of SMAD-2/3 signaling promoted derivation of MSCs from human embryonic stem cells (ESCs) rather than human iPSCs. This method was modified by using chemically de-fined mTeSR1 medium [127]. It was supplemented with SMAD-2/3 inhibitor (SB-431542) and an atmosphere of 7.5% CO_2 [128] to culture cell colonies on Matrigel-coated plates. This modified method could derive MSCs efficiently from human iPSCs, which was shown by flow cytometric analysis. MSC markers were expressed by the vast majority of adherent cells (>99.6% for CD73, CD105, and CD166 and >88.4% for CD44 and CD90). MSCs derived from transgene-free human iPSCs (iPS-MSCs) were readily expandable. And under oncogenic conditions, iPS-MSCs showed a lower potential for promoting epithelial-mesenchymal transition, invasion, stemness and growth of cancer cells [125].

An interesting study compared the ability of iPS-MSCs and primary MSCs to act as feeder cells supporting the growth of hematopoietic progenitor cells (HPCs). It was shown that iPSCs could support HPC proliferation and maintain their immunophenotype and colony-forming unit (CFU) potential. However, in a long-term study, the expression of vascular cell adhesion molecule 1 (VCAM-1) was down-regulated in iPS-MSCs compared to primary MSCs. It may be correlated with the lower long-term culture-initiating cell (LTC-IC) frequency. In other words, iPSCs were not as suitable as primary MSCs for supporting HPC culture [129].

Along with their capacity to differentiate and transdifferentiate into cells of different lineages, iPS-MSCs have also generated great interest for their ability to display immunomodulatory capacities [130]. Frobel *et al.* derived iPSCs from human BMSCs and then redifferentiated them into MSCs. The primary MSCs and iPS-MSCs showed similar gene expression and DNA methylation profiles whereas primary MSCs expressed more T-cell activation and immune response-associated proteins. Thus, it could mean that they have better immunomodulatory properties than iPS-MSCs [131]. Mesenchymal stem cells have been used in regenerative medicine to treat a number of diseases including cardiovascular disease. Mesenchymal stem cells act as a repair cell that is stimulated by physiological need. Chronic inflammation plays an integral role in the cascade leading to heart failure, and mesenchymal stem cells may be further developed to function as a biological anti-inflammatory factor [132].

Neural Stem/Progenitor Cells Derived from Induced Pluripotent Stem Cells

It has been reported that induced neural stem cells (iNSCs) can be obtained from rodent and human somatic cells through the forced expression of defined factors. Neural stem cells can be used to treat nervous system diseases, such as Parkinson's disease, amyotrophic lateral sclerosis, Alzheimer's disease and spinal cord injury. Two different approaches have been successfully used to obtain iNSCs; a direct and an indirect method, which involve an unstable intermediate state.

iPSCs could also be reprogrammed from live cells in human urine (HUC) [132]. HUCs were transfected with non-integrating oriP/EBNA episomal vectors carrying Oct4, Sox2, SV40LT, Klf4 and microRNA MIR302-367. It was done *via* electrophoretic transfer, followed by the addition of FGF-2 and five types of small molecules into the culture medium (including CHIR99021, PD0325901, A83-01, thiazovivin, and DMH1). Transfected cells showed the rosette-like morphology of typical neuroprogenitor cells (NPCs). These obtained cells could proliferate *in vitro* and differentiate into neuronal subtypes and astrocytes [133]. There were visible neural stem cells cloning twelve days after electrophoretic transfer. And they could express neural stem cell signature genes Sox1, Sox2 and PAX6.

Induced neural stem cells can differentiate into astrocytes, dopaminergic neurons, and glutamatergic neurons, but cannot differentiate into oligodendrocytes. However, after using small stimulation molecule PDGF - AA or NT3, they could differentiate into oligodendrocytes [134]. At the same time, after transplanting the induced neural stem cells into the striatum of newborn mice, neural stem cells could survive and migrate without brain tumor formation. It proves that induced neural stem cells have differentiation capacity *in vitro*. Oct4, Sox2, Klf4 and c-Myc were transferred into human fibroblasts using the Sendai virus, three small molecules, LIF, CHIR99021, and SB431542. They were added to the neural stem cell culture medium; and the temperature was elevated (39 °C) to inactivate the sendai virus. Thirteen days later, visible neural stem cell clones had formed. Those cells express neural stem cell signature genes (Sox1, Sox2, FABP7, PAX6, HES5 and NOTCH1) and can differentiate into astrocytes, neurons and oligodendrocytes *in vitro*.

Several animal studies have evaluated the efficacy of transplanting iPS-NPCs in the treatment of spinal cord injury (SCI), but most of them just delivered the cells directly to the affected area where most of the transplanted cells cannot survive for a long time [135]. However, iPS-NPC transplantation improved the function of the injured spinal cord by forming synapses with the host tissue [136] and remyelinating the injured axons [137]. In addition, the transplanted iPS-NPCs could also protect the injured cord from secondary damage through immunoregulation and neurotrophic effects [138]. However, in a rat model, most of the grafted cells underwent neural and astroglial differentiations, while a small portion of them retained stemness and kept proliferating at the end point. These uncontrolled cells pose a potential tumor formation risk, which could explain the functional decrease shown in the experimental animals [139].

Other Differentiation Potentials of Induced Pluripotent Stem Cells

In 2008, Mauritz *et al.* [140] and Narazaki *et al.* [141] proved that iPSCs can differentiate into cardiomyocytes *in vitro*, but their differentiation efficiency is lower than mESCs. These studies not only provided affirmation for iPSCs as a new source for myocardial cell transplants, but also further proved iPSC pluripotency and differentiation ability [142].

Dimos *et al.* derived iPSCs from dermal fibroblasts of an 82-year-old woman diagnosed with amyotrophic lateral sclerosis (ALS). They directly induced them to differentiate into motor neurons. This study demonstrated the feasibility of deriving large amounts of motor neurons from specific aged patients, which could be exactly immune-matched to that individual. However, a mechanism for correcting the intrinsic defects of the derived cells should be further investigated [143].

Many studies have shown pancreatic cells, which are the insulin-secreting cells, could also be derived from iPSCs by mimicking the natural developmental processes. The stability of insulin secretion of the induced cells should be an important concern. Yabe developed a 6-stage protocol to generate pancreatic β cells by adding CHIR99021 (a selective inhibitor of GSK-3β) and activin, FGF2, and BMP4 during the definitive endodermal induction. Their results showed that the induced cells exhibited more efficient insulin secretion compared to monolayer-cultured cells, and they achieved satisfactory lowered blood glucose results in a mouse model. The authors concluded that the induction of definitive endoderm and spheroid formation are the key steps for the entire process. Transplanting iPSC-pancreatic β cells provides a new option for future diabetes treatment [144].

For the treatment of many liver diseases, hepatocyte transplantation could be used as an alternative to liver transplantation. However, primary hepatocytes rapidly lose their function in culture, and the viability after cryopreservation also varies widely [145]. Therefore, finding other sources of hepatocytes seems essential, and iPSCs could be adopted as a good candidate. To obtain hepatocyte-like cells (iPSC-HLCs), the iPSCs have to experience 3 stages: definitive endoderm, hepatic progenitors and mature hepatocyte-like cells. Several protocols have been developed to induce iPSC hepatic differentiation [146, 147]. In general, the exposure of iPSCs to activin A and BMP 4, with a short exposure to Wnt3a and FGF in the first stage, is important for the fate decision of iPSC differentiation [145]. Recent studies characterized the human iPSC-HLC more like fetal hepatocytes rather than adult hepatocytes. To overcome this issue, Nakamori *et al.* transduced three transcription factors (ATF5, c/EBPα, and PROX1), into human iPSC-HLCs. It enhanced their expression and resulted in modified iPSC-HLCs, similar to the adult hepatocytes [148].

CANCER STEM CELLS AND TISSUE ENGINEERING SKILLS IN CANCER RESEARCH

Cancer has become one of the most devastating diseases in the world, with only a modest improvement in 5-year survival rates of patients [149 - 151]. It is increasingly becoming clear that cancer is a stem cell disorder. Further research is necessary to look at the similarities and differences between normal and cancer stem cells self-renewal processes [149]. The prevalence and morbidity of cancer is accompanied by the need for reconstructive procedures to minimize the disfigurement and loss of function.

One of the most enlightening discoveries in the treatment of cancer in recent years has been the tissue engineering (TE) modality. TE has evoked new concepts for the cure of organ failure and tissue loss by creating functional substitutes. For example, TE can play a major role in the rehabilitation of cancer patients following surgery [152 - 155]. In addition, the combination of TE with innovative methods of molecular biology and stem cell technology may help in exploring treatment options. It also provides insights to investigate and potentially modulate principal phenomena of tumor growth and invasion, as well as the platforms of tumor-related angiogenesis [156].

Cancer Stem Cells

Cancer stem cells (CSCs) are thought to be responsible for chemotherapy resistance and relapse after cancer treatment [157]. They have the same functional properties as normal stem cells. For instance, self-renewal, multi-potent differentiation, and the capacity to generate a heterogeneous lineage of all types of cancer cells [158]. CSCs are defined operationally by their capacity to (xeno) transplant a tumor to generate a phenocopy of the original human malignancy. Compared with the more differentiated tumor cells, CSCs are thus suggested to embody the driving force within a tumor [159].

CSCs were primarily documented for leukemia and multiple myeloma as a small subset of cancer cells that are capable of extensive proliferation [150]. With in-depth research, an increasing number of both solid and lymphoid malignancies have also been shown to contain CSCs [159]. CSCs have been shown to be key to tumor promotion in several tumor types [151]. The properties of CSCs appear to be influenced by both the specific genetic aberrations, in a given tumor, and the stage of disease progression. Additionally, the types of drugs used to challenge the tumor growth could modify the cellular properties [149, 151, 160]. The tissue microenvironment determines the fate of cancerous cells [161]. Therefore, an engineered synthetic tissue can be designed to isolate and examine the effects of individual factors in the tumor microenvironment on tumorigenesis. The growth and suppression of CSCs can also be examined on engineered dishes [157, 162, 163]. It has been suggested that blood vessels provide a niche that maintain stemness in normal organs. Identifying the CSCs and their niche are critical for identifying molecular targets to inhibit their growth and eliminate their niche [164 - 166].

Tissue Engineering Tools in Cancer

TE technologies have been designed to reconstitute damaged tissue structure and function. TE tools provide a setup that mimic tissue physiological regeneration processes and cancer pathological development as well. Such a field enables three-dimensional (3D) design of biomaterials and scaffolds to recreate the geometry, chemistry, function and signaling milieu of the native tumor microenvironment [167]. TE-cancer research models allow us to investigate various interactions and replicate physiological and pathological conditions to possibly aid in treatment [156].

Three-dimensional Tissue Culture Skills

In the past, cancer-related studies were performed in two-dimensional (2D) monolayer cultures, small animal models, and tissue specimens of human tumors. When cultured *in vitro*, cancer cells lose many of their *in vivo* features due to the absence of environmental signals present in native tumors. As a result, some important characteristics of cancer cannot be properly assessed. Animal models have some limitations as well, as they often fail to represent the pathological nature of human tumors. Bioengineering methods that have transformed stem cell research and applications in regenerative medicine are just starting to solve such problems [168]. Incorporating cancer cells in the 3D mimicking environment is promising for maintaining *in vivo* cancer behaviors in spatial and temporal contexts [155]. It became possible to observe and model tumor behavior under different conditions in a 3D construct. And the performance of antiangiogenic drugs or nanomaterials can be assessed [169].

The most widely used 3D model is the human tumor spheroid. It is a small, tightly bound cellular aggregate that tends to form when transformed cells are maintained under non-adherent conditions [170]. It is commonly used because of the 3D architecture. The extensive cell-to-cell contacts provided by spheroid growth appear to better mimic the *in vivo* cellular environment. Many of the biological properties of solid tumors, including cell morphology and characteristics, could also be simulated [171]. One of the major uses of human tumor spheroids is as a tool for preclinical screening and testing. For example, anticancer drugs, cell/antibody-based therapies and various experimental delivery and therapeutic systems [170]. Further integration of biological principles and engineering approaches will improve the understanding of cancer progression, invasion and spreading. Such advancement will support the discovery of more personalized therapies for cancer patients [172].

Microfluidics for Engineering Tumors

Biomimetic models of human tumors that provide the appropriate conditions for cancer cells are now becoming an important apparatus to investigate various tumors. Such models should have the essential factors necessary for studying specific aspects of tumor biology and pathophysiology [173]. Microfluidics is a multidisciplinary field intersecting engineering, physics, chemistry, biochemistry, nanotechnology, and biotechnology. Microfluidics has practical applications for the design of systems in which low volumes of fluids are processed to achieve multiplexing, automation, and high-throughput screening [174, 175]. It emerged in the beginning of the 1980s and addresses the behavior, precise control and manipulation of fluids that are geometrically constrained to a small, typically sub-millimeter scale. Microfluidics has been evolving to offer an opportunity to reveal critical parameters and insights that help with the translation of pharmacological advances to clinical trials [176, 177]. Additionally, microfluidics holds a promising value in diagnosing and understanding cancer biology [178] due to its high sensitivity, high throughput, and enhanced spatio-temporal control, low cost and less material-consumption. Furthermore, microfluidic-based platforms are portable and can be easily designed for point-of-care diagnostics [179].

Nanotechnology

Nanomaterials are increasingly being investigated for biomedical applications, especially in cancer theranostics and tissue engineering [168, 180]. Targeting nanoparticles to tumor microenvironments could overcome several biological barriers that could hinder nanoparticles from performing their therapeutic functions. Parameters such as composition, surface modification and particle size affect a wide range of biomaterial properties. These paprameters result in different biodistribution profiles and delivery depth. The nanosystems have been successfully applied to tumor detection and anti-tumor drug delivery [181]. The size, shape and surface modification of the nanoparticles are critical for the investigation of nanotoxicology and biocompatibility in a 3D cell culture construct. The dynamic conditions such as flow and mechanical stimulation, as well as physiological movements, significantly affect the toxicity induced by nanoparticles [169]. Recent advances in nanotechnology have contributed to the development of engineered nanoscale materials as innovative prototypes to be used for biomedical applications and optimized therapy [27]. Nanoparticles can be selectively developed to deliver bioactive agents to specific biological targets. It is due to their unique features such as a large specific surface area, structural properties, and long circulation time in the blood [182]. Using nanotechnology to prevent and target cancerous cells can facilitate site-selective drug delivery, increase drug accumulation in the tumor site and consequently reduce side effects [183, 184].

CONCLUDING REMARKS AND FUTURE PERSPECTIVES

Current human stem cell research has generated substantial achievements in isolating stem/progenitor cells in almost all organs. Advancement in *in vitro* tools has been providing a better understanding of characterizing the mechanisms of self-renewal, proliferation and differentiation. One of these *in vitro*-based outcomes is the generation of iPSCs, which could be considered a vital breakthrough of this decade. It provides a possibility for developing patient-specific stem cells, which could act as the "integrator" of precision medicine. However, barriers regarding safety and efficiency remain unsolved prior to translating stem cell research into clinical practice. In particular, we expect to see more promising results with high throughput. One of which is a non-virus reprograming of terminal differentiated cells to iPS cells and then, directing their differentiation into designed cells. Another expectation could be on developing robust *in vitro* culture systems (bioreactors) to effectively expand stem cells and monitor their metabolism and phenotype to generate sufficient cells for clinical use.

Tissue engineering methods have also transformed stem cell research to include the field of cancer investigation. Cancer engineering is a relatively recent field and has many challenges to overcome before being applied to screening and testing of cancer drugs. The ongoing developments in human tumor platforms can be a prelude for personalized medicine. It is to evaluate the tumor progression for the patients using their own tumor cells. The patient-tailored platforms could be used to explore optimal and individualized drug delivery.

ABBREVIATIONS

AHF	acute hepatic failure
ASCs	adipose derived stem cells
BDNF	brain-derived neurotrophic factor
BMP	bone morphogenetic protein
CNS	central nervous system
CPCs	cardiogenic progenitor cells
CSCs	cardiac stem cells
DFPCs	dental follicle progenitor cells
DG	dentate gyrus
DPSCs	dental pulp stem cells
EC	embryonal carcinoma
EGF	epidermal growth factor
ESCs	embryonic stem cells
FGF	fibroblast growth factor
GABA	γ-aminobutyric acid
G-CSF	granulocyte-colony stimulating factor
HGF	hepatocyte growth factor
IGF	insulin-like growth factor
iNSCs	induced neural stem cells
iPSCs	induced pluripotent stem cells
LIF	leukemia inhibiting factor
LPCs	liver stem/progenitor cells
MEFs	mouse embryonic fibroblasts
MI	myocardial infarction
MSCs	mesenchymal stem cells
NGF	nerve growth factor
NO	nitric oxide
NSCs	neural stem cells
OB	olfactory bulb
PDGF	platelet-derived growth factor
PDLSCs	periodontal ligament stem cells
Sca	stem cell antigen
SCAP	stem cells from apical papilla
SCNT	somatic cell nuclear transfer

SGZ	subgranular zone
SSCs	somatic stem cells
SVZ	subventricular zone
SHED	stem cells from exfoliated deciduous teeth
TNF	tumor necrosis factor
VEGF	vascular endothelial growth factor

CONFLICT OF INTEREST

The authors declare no conflict of interest, financial or otherwise.

ACKNOWLEDGEMENTS

Declared none.

REFERENCES

[1]   Stevens LC, Little CC. Spontaneous testicular teratomas in an inbred strain of mice. Proc Natl Acad Sci USA 1954; 40(11): 1080-7.
[http://dx.doi.org/10.1073/pnas.40.11.1080] [PMID: 16578442]

[2]   Kleinsmith LJ, Pierce GB Jr. Multipotentiality of single embryonal carcinoma cells. Cancer Res 1964; 24: 1544-51.
[PMID: 14234000]

[3]   Kahan BW, Ephrussi B. Developmental potentialities of clonal *in vitro* cultures of mouse testicular teratoma. J Natl Cancer Inst 1970; 44(5): 1015-36.
[PMID: 5514468]

[4]   James AT, Joseph IE, Sander SS, *et al.* Embryonic Stem Cell Lines Derived from Human Blastocysts. Science 1998; 6;282(5391): 1145-7.

[5]   Shroff G, Gupta R. Human embryonic stem cells in the treatment of patients with spinal cord injury. Ann Neurosci 2015; 22(4): 208-16.
[http://dx.doi.org/10.5214/ans.0972.7531.220404] [PMID: 26526627]

[6]   Ambasudhan R, Dolatabadi N, Nutter A, Masliah E, Mckercher SR, Lipton SA. Potential for cell therapy in Parkinson's disease using genetically programmed human embryonic stem cell-derived neural progenitor cells. J Comp Neurol 2014; 522(12): 2845-56.
[http://dx.doi.org/10.1002/cne.23617] [PMID: 24756727]

[7]   Reynolds J, Lamba DA. Human embryonic stem cell applications for retinal degenerations. Exp Eye Res 2014; 123: 151-60.
[http://dx.doi.org/10.1016/j.exer.2013.07.010] [PMID: 23880530]

[8]   Atkinson SP, Lako M, Armstrong L. Potential for pharmacological manipulation of human embryonic stem cells. Br J Pharmacol 2013; 169(2): 269-89.
[http://dx.doi.org/10.1111/j.1476-5381.2012.01978.x] [PMID: 22515554]

[9]   Resnick JL, Bixler LS, Cheng L, Donovan PJ. Long-term proliferation of mouse primordial germ cells in culture. Nature 1992; 359(6395): 550-1.
[http://dx.doi.org/10.1038/359550a0] [PMID: 1383830]

[10]   Matsui Y, Zsebo K, Hogan BL. Derivation of pluripotential embryonic stem cells from murine primordial germ cells in culture. Cell 1992; 70(5): 841-7.

[http://dx.doi.org/10.1016/0092-8674(92)90317-6] [PMID: 1381289]

[11]   Trounson AO. The derivation and potential use of human embryonic stem cells. Reprod Fertil Dev 2001; 13(7-8): 523-32.
[http://dx.doi.org/10.1071/RD01101] [PMID: 11999302]

[12]   Trounson A. The production and directed differentiation of human embryonic stem cells. Endocr Rev 2006; 27(2): 208-19.
[http://dx.doi.org/10.1210/er.2005-0016] [PMID: 16434509]

[13]   Kraushaar DC, Dalton S, Wang L. Heparan sulfate: a key regulator of embryonic stem cell fate. 2013; 394(6): 741-51.
[http://dx.doi.org/10.1515/hsz-2012-0353]

[14]   Wang H, Luo X, Leighton J. Extracellular Matrix and Integrins in Embryonic Stem Cell Differentiation. Biochem Insights 2015; 8 (Suppl. 2): 15-21.
[http://dx.doi.org/10.4137/BCI.S30377] [PMID: 26462244]

[15]   Pera MF, Andrade J, Houssami S, et al. Regulation of human embryonic stem cell differentiation by BMP-2 and its antagonist noggin. J Cell Sci 2004; 117(Pt 7): 1269-80.
[http://dx.doi.org/10.1242/jcs.00970] [PMID: 14996946]

[16]   Perrier AL, Tabar V, Barberi T, et al. Derivation of midbrain dopamine neurons from human embryonic stem cells. Proc Natl Acad Sci USA 2004; 101(34): 12543-8.
[http://dx.doi.org/10.1073/pnas.0404700101] [PMID: 15310843]

[17]   Magli A, Schnettler E, Swanson SA, et al. Pax3 and Tbx5 specify whether PDGFRα+ cells assume skeletal or cardiac muscle fate in differentiating embryonic stem cells. Stem Cells 2014; 32(8): 2072-83.
[http://dx.doi.org/10.1002/stem.1713] [PMID: 24677751]

[18]   Wong MM, Yin X, Potter C, et al. Over-expression of HSP47 augments mouse embryonic stem cell smooth muscle differentiation and chemotaxis. PLoS One 2014; 9(1): e86118.
[http://dx.doi.org/10.1371/journal.pone.0086118] [PMID: 24454956]

[19]   Cheung C, Sinha S. Human embryonic stem cell-derived vascular smooth muscle cells in therapeutic neovascularisation. J Mol Cell Cardiol 2011; 51(5): 651-64.
[http://dx.doi.org/10.1016/j.yjmcc.2011.07.014] [PMID: 21816157]

[20]   Cerdan C, Rouleau A, Bhatia M. VEGF-A165 augments erythropoietic development from human embryonic stem cells. Blood 2004; 103(7): 2504-12.
[http://dx.doi.org/10.1182/blood-2003-07-2563] [PMID: 14656883]

[21]   Kawaguchi M, Kitajima K, Kanokoda M, et al. Efficient production of platelets from mouse embryonic stem cells by enforced expression of Gata2 in late hemogenic endothelial cells. Biochem Biophys Res Commun 2016; 474(3): 462-8.
[http://dx.doi.org/10.1016/j.bbrc.2016.04.140] [PMID: 27131743]

[22]   Ding H, Xu X, Qin X, Yang C, Feng Q. Resveratrol promotes differentiation of mouse embryonic stem cells to cardiomyocytes. Cardiovasc Ther 2016; 34(4): 283-9.
[http://dx.doi.org/10.1111/1755-5922.12200] [PMID: 27225714]

[23]   Ishii T, Yasuchika K, Ikai I. Hepatic differentiation of embryonic stem cells by murine fetal liver mesenchymal cells. Methods Mol Biol 2013; 946: 469-78.
[http://dx.doi.org/10.1007/978-1-62703-128-8_29] [PMID: 23179850]

[24]   Brolén GK, Heins N, Edsbagge J, Semb H. Signals from the embryonic mouse pancreas induce differentiation of human embryonic stem cells into insulin-producing beta-cell-like cells. Diabetes 2005; 54(10): 2867-74.
[http://dx.doi.org/10.2337/diabetes.54.10.2867] [PMID: 16186387]

[25]   Candiello J, Singh SS, Task K, Kumta PN, Banerjee I. Early differentiation patterning of mouse embryonic stem cells in response to variations in alginate substrate stiffness. J Biol Eng 2013; 7(1): 9.

[http://dx.doi.org/10.1186/1754-1611-7-9] [PMID: 23570553]

[26]   Scott CT, McCormick JB, Owen-Smith J. And then there were two: use of hESC lines. Nat Biotechnol 2009; 27(8): 696-7.
[http://dx.doi.org/10.1038/nbt0809-696] [PMID: 19668169]

[27]   Fraga AM, Souza de Araújo ÉS, Stabellini R, Vergani N, Pereira LV. A survey of parameters involved in the establishment of new lines of human embryonic stem cells. Stem Cell Rev 2011; 7(4): 775-81.
[http://dx.doi.org/10.1007/s12015-011-9250-x] [PMID: 21416256]

[28]   Daley GQ. Missed opportunities in embryonic stem-cell research. N Engl J Med 2004; 351(7): 627-8.
[http://dx.doi.org/10.1056/NEJMp048200] [PMID: 15302910]

[29]   Modliński JA. Haploid mouse embryos obtained by microsurgical removal of one pronucleus. J Embryol Exp Morphol 1975; 33(4): 897-905.
[PMID: 1176880]

[30]   Tarkowski AK, Rossant J. Haploid mouse blastocysts developed from bisected zygotes. Nature 1976; 259(5545): 663-5.
[http://dx.doi.org/10.1038/259663a0] [PMID: 1250417]

[31]   Kaufman MH. Chromosome analysis of early postimplantation presumptive haploid parthenogenetic mouse embryos. J Embryol Exp Morphol 1978; 45: 85-91.
[PMID: 670867]

[32]   Latham KE, Akutsu H, Patel B, Yanagimachi R. Comparison of gene expression during preimplantation development between diploid and haploid mouse embryos. Biol Reprod 2002; 67(2): 386-92.
[http://dx.doi.org/10.1095/biolreprod67.2.386] [PMID: 12135871]

[33]   Yang H, Shi L, Wang BA, *et al.* Generation of genetically modified mice by oocyte injection of androgenetic haploid embryonic stem cells. Cell 2012; 149(3): 605-17.
[http://dx.doi.org/10.1016/j.cell.2012.04.002] [PMID: 22541431]

[34]   Egli D, Chen AE, Saphier G, *et al.* Reprogramming within hours following nuclear transfer into mouse but not human zygotes. Nat Commun 2011; 2: 488.
[http://dx.doi.org/10.1038/ncomms1503] [PMID: 21971503]

[35]   Noggle S, Fung HL, Gore A, *et al.* Human oocytes reprogram somatic cells to a pluripotent state. Nature 201;478(7367): 70-5.
[http://dx.doi.org/10.1038/nature10397]

[36]   Tachibana M, Amato P, Sparman M, *et al.* Human embryonic stem cells derived by somatic cell nuclear transfer. Cell 2013; 153(6): 1228-38.
[http://dx.doi.org/10.1016/j.cell.2013.05.006] [PMID: 23683578]

[37]   Cyranoski D. Human stem cells created by cloning. Nature 2013; 497(7449): 295-6.
[http://dx.doi.org/10.1038/497295a] [PMID: 23676729]

[38]   Puri MC, Nagy A. Concise review: Embryonic stem cells *versus* induced pluripotent stem cells: the game is on. Stem Cells 2012; 30(1): 10-4.
[http://dx.doi.org/10.1002/stem.788] [PMID: 22102565]

[39]   Choi J, Lee S, Mallard W, *et al.* A comparison of genetically matched cell lines reveals the equivalence of human iPSCs and ESCs. Nat Biotechnol 2015; 33(11): 1173-81.
[http://dx.doi.org/10.1038/nbt.3388] [PMID: 26501951]

[40]   Schulz TC. Concise Review: Manufacturing of pancreatic endoderm cells for clinical trials in type 1 diabetes. Stem Cells Transl Med 2015; 4(8): 927-31.
[http://dx.doi.org/10.5966/sctm.2015-0058] [PMID: 26062982]

[41]   Giakoumopoulos M, Golos TG. Embryonic stem cell-derived trophoblast differentiation: a comparative review of the biology, function, and signaling mechanisms. J Endocrinol 2013; 216(3):

R33-45.
[http://dx.doi.org/10.1530/JOE-12-0433] [PMID: 23291503]

[42] Singh VK, Kalsan M, Kumar N, Saini A, Chandra R. Induced pluripotent stem cells: applications in regenerative medicine, disease modeling, and drug discovery. Front Cell Dev Biol 2015; 3: 2.
[http://dx.doi.org/10.3389/fcell.2015.00002] [PMID: 25699255]

[43] Gokhale PJ, Andrews PW. The development of pluripotent stem cells. Curr Opin Genet Dev 2012; 22(5): 403-8.
[http://dx.doi.org/10.1016/j.gde.2012.07.006] [PMID: 22868175]

[44] Mayshar Y, Ben-David U, Lavon N, *et al.* Identification and classification of chromosomal aberrations in human induced pluripotent stem cells. Cell Stem Cell 2010; 7(4): 521-31.
[http://dx.doi.org/10.1016/j.stem.2010.07.017] [PMID: 20887957]

[45] Hussein SM, Batada NN, Vuoristo S, Ching RW, Autio R, Närvä E. Copy number variation and selection during reprogramming to pluripotency. Nature 2011; 3;471(7336): 58-62.

[46] Laurent LC, Ulitsky I, Slavin I, *et al.* Dynamic changes in the copy number of pluripotency and cell proliferation genes in human ESCs and iPSCs during reprogramming and time in culture. Cell Stem Cell 2011; 8(1): 106-18.
[http://dx.doi.org/10.1016/j.stem.2010.12.003] [PMID: 21211785]

[47] Lee JH, Lee JB, Shapovalova Z, *et al.* Somatic transcriptome priming gates lineage-specific differentiation potential of human-induced pluripotent stem cell states. Nat Commun 2014; 5: 5605.
[http://dx.doi.org/10.1038/ncomms6605] [PMID: 25465724]

[48] Goodell MA, Nguyen H, Shroyer N. Somatic stem cell heterogeneity: diversity in the blood, skin and intestinal stem cell compartments. Nat Rev Mol Cell Biol 2015; 16(5): 299-309.
[http://dx.doi.org/10.1038/nrm3980] [PMID: 25907613]

[49] O'Connor TP, Crystal RG. Genetic medicines: treatment strategies for hereditary disorders. Nat Rev Genet 2006; 7(4): 261-76.
[http://dx.doi.org/10.1038/nrg1829] [PMID: 16543931]

[50] Friedenstein AJ, Chailakhjan RK, Lalykina KS. The development of fibroblast colonies in monolayer cultures of guinea-pig bone marrow and spleen cells. Cell Tissue Kinet 1970; 3(4): 393-403.
[PMID: 5523063]

[51] Zuk PA, Zhu M, Mizuno H, *et al.* Multilineage cells from human adipose tissue: implications for cell-based therapies. Tissue Eng 2001; 7(2): 211-28.
[http://dx.doi.org/10.1089/107632701300062859] [PMID: 11304456]

[52] Alsalameh S, Amin R, Gemba T, Lotz M. Identification of mesenchymal progenitor cells in normal and osteoarthritic human articular cartilage. Arthritis Rheum 2004; 50(5): 1522-32.
[http://dx.doi.org/10.1002/art.20269] [PMID: 15146422]

[53] Trivanović D, Kocić J, Mojsilović S, *et al.* Mesenchymal stem cells isolated from peripheral blood and umbilical cord Wharton's jelly. Srp Arh Celok Lek 2013; 141(3-4): 178-86.
[http://dx.doi.org/10.2298/SARH1304178T] [PMID: 23745340]

[54] Lee OK, Kuo TK, Chen WM, Lee KD, Hsieh SL, Chen TH. Isolation of multipotent mesenchymal stem cells from umbilical cord blood. Blood 2004; 103(5): 1669-75.
[http://dx.doi.org/10.1182/blood-2003-05-1670] [PMID: 14576065]

[55] Li CD, Zhang WY, Li HL, *et al.* Mesenchymal stem cells derived from human placenta suppress allogeneic umbilical cord blood lymphocyte proliferation. Cell Res 2005; 15(7): 539-47.
[http://dx.doi.org/10.1038/sj.cr.7290323] [PMID: 16045817]

[56] De Bari C, Dell'Accio F, Tylzanowski P, Luyten FP. Multipotent mesenchymal stem cells from adult human synovial membrane. Arthritis Rheum 2001; 44(8): 1928-42.
[http://dx.doi.org/10.1002/1529-0131(200108)44:8<1928::AID-ART331>3.0.CO;2-P]   [PMID: 11508446]

[57]   Wei X, Yang X, Han ZP, Qu FF, Shao L, Shi YF. Mesenchymal stem cells: a new trend for cell therapy. Acta Pharmacol Sin 2013; 34(6): 747-54.
[http://dx.doi.org/10.1038/aps.2013.50] [PMID: 23736003]

[58]   Nombela-Arrieta C, Ritz J, Silberstein LE. The elusive nature and function of mesenchymal stem cells. Nat Rev Mol Cell Biol 2011; 12(2): 126-31.
[http://dx.doi.org/10.1038/nrm3049] [PMID: 21253000]

[59]   Pittenger MF, Mackay AM, Beck SC, *et al.* Multilineage potential of adult human mesenchymal stem cells. Science 1999; 284(5411): 143-7.
[http://dx.doi.org/10.1126/science.284.5411.143] [PMID: 10102814]

[60]   Dominici M, Le Blanc K, Mueller I, *et al.* Minimal criteria for defining multipotent mesenchymal stromal cells. The International Society for Cellular Therapy position statement. Cytotherapy 2006; 8(4): 315-7.
[http://dx.doi.org/10.1080/14653240600855905] [PMID: 16923606]

[61]   Smart I. The subependymal layer of the mouse brain and its cell production as shown by radioautography after thymidine-H3 injection. J Comp Neurol 1961; 116: 325-47.
[http://dx.doi.org/10.1002/cne.901160306]

[62]   Díez-Tejedor E, Gutiérrez-Fernández M, Martínez-Sánchez P, *et al.* Reparative therapy for acute ischemic stroke with allogeneic mesenchymal stem cells from adipose tissue: a safety assessment: a phase II randomized, double-blind, placebo-controlled, single-center, pilot clinical trial. J Stroke Cerebrovasc Dis 2014; 23(10): 2694-700.
[http://dx.doi.org/10.1016/j.jstrokecerebrovasdis.2014.06.011] [PMID: 25304723]

[63]   Vega A, Martín-Ferrero MA, Del Canto F, *et al.* Treatment of Knee Osteoarthritis With Allogeneic Bone Marrow Mesenchymal Stem Cells: A Randomized Controlled Trial. Transplantation 2015; 99(8): 1681-90.
[http://dx.doi.org/10.1097/TP.0000000000000678] [PMID: 25822648]

[64]   Mathiasen AB, Haack-Sørensen M, Jørgensen E, Kastrup J. Autotransplantation of mesenchymal stromal cells from bone-marrow to heart in patients with severe stable coronary artery disease and refractory angina--final 3-year follow-up. Int J Cardiol 2013; 170(2): 246-51.
[http://dx.doi.org/10.1016/j.ijcard.2013.10.079] [PMID: 24211066]

[65]   Galli R, Gritti A, Bonfanti L, Vescovi AL. Neural stem cells: an overview. Circ Res 2003; 92(6): 598-608.
[http://dx.doi.org/10.1161/01.RES.0000065580.02404.F4] [PMID: 12676811]

[66]   Braun SM, Jessberger S. Adult neurogenesis and its role in neuropsychiatric disease, brain repair and normal brain function. Neuropathol Appl Neurobiol 2014; 40(1): 3-12.
[http://dx.doi.org/10.1111/nan.12107] [PMID: 24308291]

[67]   Rietze R, Poulin P, Weiss S. Mitotically active cells that generate neurons and astrocytes are present in multiple regions of the adult mouse hippocampus. J Comp Neurol 2000; 424(3): 397-408.
[http://dx.doi.org/10.1002/1096-9861(20000828)424:3<397::AID-CNE2>3.0.CO;2-A]   [PMID: 10906710]

[68]   Merkle FT, Mirzadeh Z, Alvarez-Buylla A. Mosaic organization of neural stem cells in the adult brain. Science 2007; 317(5836): 381-4.
[http://dx.doi.org/10.1126/science.1144914] [PMID: 17615304]

[69]   Urbán N, Guillemot F. Neurogenesis in the embryonic and adult brain: same regulators, different roles. Front Cell Neurosci 2014; 8: 396.

[70]   Basak O, Taylor V. Stem cells of the adult mammalian brain and their niche. Cell Mol Life Sci 2009; 66(6): 1057-72.
[http://dx.doi.org/10.1007/s00018-008-8544-x] [PMID: 19011753]

[71]   He Y, Zhang PZ, Sun D, *et al.* Wnt1 from cochlear schwann cells enhances neuronal differentiation of

transplanted neural stem cells in a rat spiral ganglion neuron degeneration model. Cell Transplant 2014; 23(6): 747-60.
[http://dx.doi.org/10.3727/096368913X669761] [PMID: 23809337]

[72]   Iwai M, Abe K, Kitagawa H, Hayashi T. Gene therapy with adenovirus-mediated glial cell line-derived neurotrophic factor and neural stem cells activation after ischemic brain injury. Hum Cell 2001; 14(1): 27-38.
[PMID: 11436351]

[73]   Alizadeh R, Hassanzadeh G, Joghataei MT, *et al. In vitro* differentiation of neural stem cells derived from human olfactory bulb into dopaminergic-like neurons. Eur J Neurosci 2017; 45(6): 773-84.
[http://dx.doi.org/10.1111/ejn.13504] [PMID: 27987378]

[74]   Marei HE, Ahmed AE. Transcription factors expressed in embryonic and adult olfactory bulb neural stem cells reveal distinct proliferation, differentiation and epigenetic control. Genomics 2013; 101(1): 12-9.
[http://dx.doi.org/10.1016/j.ygeno.2012.09.006] [PMID: 23041222]

[75]   Marei HE, Althani A, Rezk S, *et al.* Therapeutic potential of human olfactory bulb neural stem cells for spinal cord injury in rats. Spinal Cord 2016; 54(10): 785-97.
[http://dx.doi.org/10.1038/sc.2016.14] [PMID: 26882489]

[76]   Berg DA, Belnoue L, Song H, Simon A. Neurotransmitter-mediated control of neurogenesis in the adult vertebrate brain. Development 2013; 140(12): 2548-61.
[http://dx.doi.org/10.1242/dev.088005] [PMID: 23715548]

[77]   Zhao C, Deng W, Gage FH. Mechanisms and functional implications of adult neurogenesis. Cell 2008; 132(4): 645-60.
[http://dx.doi.org/10.1016/j.cell.2008.01.033] [PMID: 18295581]

[78]   Katz B. The Terminations of the Afferent Nerve Fibre in the Muscle Spindle of the Frog. Philos Trans R Soc Lond B Biol Sci 1961; 243: 221-40.
[http://dx.doi.org/10.1098/rstb.1961.0001]

[79]   Mauro A. Satellite cell of skeletal muscle fibers. J Biophys Biochem Cytol 1961; 9: 493-5.
[http://dx.doi.org/10.1083/jcb.9.2.493] [PMID: 13768451]

[80]   Hawke TJ, Garry DJ. Myogenic satellite cells: physiology to molecular biology. J Appl Physiol 2001; 91(2): 534-51.
[PMID: 11457764]

[81]   Yin H, Price F, Rudnicki MA. Satellite cells and the muscle stem cell niche. Physiol Rev 2013; 93(1): 23-67.
[http://dx.doi.org/10.1152/physrev.00043.2011] [PMID: 23303905]

[82]   Morgan JE, Partridge TA. Muscle satellite cells. Int J Biochem Cell Biol 2003; 35(8): 1151-6.
[http://dx.doi.org/10.1016/S1357-2725(03)00042-6] [PMID: 12757751]

[83]   Collins CA, Olsen I, Zammit PS, *et al.* Stem cell function, self-renewal, and behavioral heterogeneity of cells from the adult muscle satellite cell niche. Cell 2005; 122(2): 289-301.
[http://dx.doi.org/10.1016/j.cell.2005.05.010] [PMID: 16051152]

[84]   Yablonka-Reuveni Z. The skeletal muscle satellite cell: still young and fascinating at 50. J Histochem Cytochem 2011; 59(12): 1041-59.
[http://dx.doi.org/10.1369/0022155411426780] [PMID: 22147605]

[85]   Beltrami AP, Barlucchi L, Torella D, *et al.* Adult cardiac stem cells are multipotent and support myocardial regeneration. Cell 2003; 114(6): 763-76.
[http://dx.doi.org/10.1016/S0092-8674(03)00687-1] [PMID: 14505575]

[86]   Bearzi C, Rota M, Hosoda T, *et al.* Human cardiac stem cells. Proc Natl Acad Sci USA 2007; 104(35): 14068-73.
[http://dx.doi.org/10.1073/pnas.0706760104] [PMID: 17709737]

[87] Urbanek K, Cesselli D, Rota M, *et al.* Stem cell niches in the adult mouse heart. Proc Natl Acad Sci USA 2006; 103(24): 9226-31.
[http://dx.doi.org/10.1073/pnas.0600635103] [PMID: 16754876]

[88] Leri A, Kajstura J, Anversa P. Role of cardiac stem cells in cardiac pathophysiology: a paradigm shift in human myocardial biology. Circ Res 2011; 109(8): 941-61.
[http://dx.doi.org/10.1161/CIRCRESAHA.111.243154] [PMID: 21960726]

[89] Sanada F, Kim J, Czarna A, *et al.* c-Kit-positive cardiac stem cells nested in hypoxic niches are activated by stem cell factor reversing the aging myopathy. Circ Res 2014; 114(1): 41-55.
[http://dx.doi.org/10.1161/CIRCRESAHA.114.302500] [PMID: 24170267]

[90] Davis DR, Stewart DJ. Selectins for cardiosphere culture: the "E's" have it! Molecular therapy : the journal of the American Society of Gene Therapy. 2012; 20: 1296-7.

[91] Sohur US, Emsley JG, Mitchell BD, Macklis JD. Adult neurogenesis and cellular brain repair with neural progenitors, precursors and stem cells. Philos Trans R Soc Lond B Biol Sci 2006; 361(1473): 1477-97.
[http://dx.doi.org/10.1098/rstb.2006.1887] [PMID: 16939970]

[92] Liu J, Wang Y, Du W, Yu B. Sca-1-positive cardiac stem cell migration in a cardiac infarction model. Inflammation 2013; 36(3): 738-49.
[http://dx.doi.org/10.1007/s10753-013-9600-8] [PMID: 23400327]

[93] Messina E, De Angelis L, Frati G, *et al.* Isolation and expansion of adult cardiac stem cells from human and murine heart. Circ Res 2004; 95(9): 911-21.
[http://dx.doi.org/10.1161/01.RES.0000147315.71699.51] [PMID: 15472116]

[94] Davis ME, Hsieh PC, Takahashi T, *et al.* Local myocardial insulin-like growth factor 1 (IGF-1) delivery with biotinylated peptide nanofibers improves cell therapy for myocardial infarction. Proc Natl Acad Sci USA 2006; 103(21): 8155-60.
[http://dx.doi.org/10.1073/pnas.0602877103] [PMID: 16698918]

[95] D'Amario D, Cabral-Da-Silva MC, Zheng H, *et al.* Insulin-like growth factor-1 receptor identifies a pool of human cardiac stem cells with superior therapeutic potential for myocardial regeneration. Circ Res 2011; 108(12): 1467-81.
[http://dx.doi.org/10.1161/CIRCRESAHA.111.240648] [PMID: 21546606]

[96] Hatzistergos KE, Quevedo H, Oskouei BN, *et al.* Bone marrow mesenchymal stem cells stimulate cardiac stem cell proliferation and differentiation. Circ Res 2010; 107(7): 913-22.
[http://dx.doi.org/10.1161/CIRCRESAHA.110.222703] [PMID: 20671238]

[97] Takehara N, Tsutsumi Y, Tateishi K, *et al.* Controlled delivery of basic fibroblast growth factor promotes human cardiosphere-derived cell engraftment to enhance cardiac repair for chronic myocardial infarction. J Am Coll Cardiol 2008; 52(23): 1858-65.
[http://dx.doi.org/10.1016/j.jacc.2008.06.052] [PMID: 19038683]

[98] Penn MS, Mangi AA. Genetic enhancement of stem cell engraftment, survival, and efficacy. Circ Res 2008; 102(12): 1471-82.
[http://dx.doi.org/10.1161/CIRCRESAHA.108.175174] [PMID: 18566313]

[99] Gaetani R, Feyen DA, Verhage V, *et al.* Epicardial application of cardiac progenitor cells in a 3D-printed gelatin/hyaluronic acid patch preserves cardiac function after myocardial infarction. Biomaterials 2015; 61: 339-48.
[http://dx.doi.org/10.1016/j.biomaterials.2015.05.005] [PMID: 26043062]

[100] Farber E. Similarities in the sequence of early histological changes induced in the liver of the rat by ethionine, 2-acetylamino-fluorene, and 3′-methyl-4-dimethylaminoazobenzene. Cancer Res 1956; 16(2): 142-8.
[PMID: 13293655]

[101] Itoh T, Miyajima A. Liver regeneration by stem/progenitor cells. Hepatology 2014; 59(4): 1617-26.

[http://dx.doi.org/10.1002/hep.26753] [PMID: 24115180]

[102]   Miyajima A, Tanaka M, Itoh T. Stem/progenitor cells in liver development, homeostasis, regeneration, and reprogramming. Cell Stem Cell 2014; 14(5): 561-74.
[http://dx.doi.org/10.1016/j.stem.2014.04.010] [PMID: 24792114]

[103]   Takase HM, Itoh T, Ino S, *et al.* FGF7 is a functional niche signal required for stimulation of adult liver progenitor cells that support liver regeneration. Genes Dev 2013; 27(2): 169-81.
[http://dx.doi.org/10.1101/gad.204776.112] [PMID: 23322300]

[104]   Jakubowski A, Ambrose C, Parr M, *et al.* TWEAK induces liver progenitor cell proliferation. J Clin Invest 2005; 115(9): 2330-40.
[http://dx.doi.org/10.1172/JCI23486] [PMID: 16110324]

[105]   Ishikawa T, Factor VM, Marquardt JU, *et al.* Hepatocyte growth factor/c-met signaling is required for stem-cell-mediated liver regeneration in mice. Hepatology 2012; 55(4): 1215-26.
[http://dx.doi.org/10.1002/hep.24796] [PMID: 22095660]

[106]   Sawitza I, Kordes C, Reister S, Häussinger D. The niche of stellate cells within rat liver. Hepatology 2009; 50(5): 1617-24.
[http://dx.doi.org/10.1002/hep.23184] [PMID: 19725107]

[107]   Hashemi Goradel N, Darabi M, Shamsasenjan K, Ejtehadifar M, Zahedi S. Methods of Liver Stem Cell Therapy in Rodents as Models of Human Liver Regeneration in Hepatic Failure. Adv Pharm Bull 2015; 5(3): 293-8.
[http://dx.doi.org/10.15171/apb.2015.041] [PMID: 26504749]

[108]   Volponi AA, Pang Y, Sharpe PT. Stem cell-based biological tooth repair and regeneration. Trends Cell Biol 2010; 20(12): 715-22.
[http://dx.doi.org/10.1016/j.tcb.2010.09.012] [PMID: 21035344]

[109]   Kim SG, Zhou J, Solomon C, *et al.* Effects of growth factors on dental stem/progenitor cells. Dent Clin North Am 2012; 56(3): 563-75.
[http://dx.doi.org/10.1016/j.cden.2012.05.001] [PMID: 22835538]

[110]   Zhang YD, Chen Z, Song YQ, Liu C, Chen YP. Making a tooth: growth factors, transcription factors, and stem cells. Cell Res 2005; 15(5): 301-16.
[http://dx.doi.org/10.1038/sj.cr.7290299] [PMID: 15916718]

[111]   Gurdon JB, Byrne JA. The first half-century of nuclear transplantation. Proc Natl Acad Sci USA 2003; 100(14): 8048-52.
[http://dx.doi.org/10.1073/pnas.1337135100] [PMID: 12821779]

[112]   Takahashi K, Yamanaka S. Induction of pluripotent stem cells from mouse embryonic and adult fibroblast cultures by defined factors. Cell 2006; 126(4): 663-76.
[http://dx.doi.org/10.1016/j.cell.2006.07.024] [PMID: 16904174]

[113]   Shi X, Lv S, He X, *et al.* Differentiation of hepatocytes from induced pluripotent stem cells derived from human hair follicle mesenchymal stem cells. Cell Tissue Res 2016; 366(1): 89-99.
[http://dx.doi.org/10.1007/s00441-016-2399-5] [PMID: 27053247]

[114]   Ulm A, Mayhew CN, Debley J, Khurana Hershey GK, Ji H. Cultivate Primary Nasal Epithelial Cells from Children and Reprogram into Induced Pluripotent Stem Cells. J Vis Exp 2016; (109):
[http://dx.doi.org/10.3791/53814] [PMID: 27022951]

[115]   Ye J, Ge J, Zhang X, *et al.* Pluripotent stem cells induced from mouse neural stem cells and small intestinal epithelial cells by small molecule compounds. Cell Res 2016; 26(1): 34-45.
[http://dx.doi.org/10.1038/cr.2015.142] [PMID: 26704449]

[116]   Unzu C, Friedli M, Bosman A, Jaconi ME, Wildhaber BE, Rougemont AL. Human Hepatocyte-Derived Induced Pluripotent Stem Cells: MYC Expression, Similarities to Human Germ Cell Tumors, and Safety Issues. Stem Cells Int 2016; 2016: 4370142.
[http://dx.doi.org/10.1155/2016/4370142] [PMID: 26880963]

[117]   Cheng H, Liu C, Cai X, Lu Y, Xu Y, Yu X. iPSCs Derived from Malignant Tumor Cells: Potential Application for Cancer Research. Curr Stem Cell Res Ther 2016; 11(5): 444-50.
[http://dx.doi.org/10.2174/1574888X11666160217154748] [PMID: 26899393]

[118]   Yoshida Y, Takahashi K, Okita K, Ichisaka T, Yamanaka S. Hypoxia enhances the generation of induced pluripotent stem cells. Cell Stem Cell 2009; 5(3): 237-41.
[http://dx.doi.org/10.1016/j.stem.2009.08.001] [PMID: 19716359]

[119]   Esteban MA, Wang T, Qin B, *et al.* Vitamin C enhances the generation of mouse and human induced pluripotent stem cells. Cell Stem Cell 2010; 6(1): 71-9.
[http://dx.doi.org/10.1016/j.stem.2009.12.001] [PMID: 20036631]

[120]   Yi L, Lu C, Hu W, Sun Y, Levine AJ. Multiple roles of p53-related pathways in somatic cell reprogramming and stem cell differentiation. Cancer Res 2012; 72(21): 5635-45.
[http://dx.doi.org/10.1158/0008-5472.CAN-12-1451] [PMID: 22964580]

[121]   Droujinine IA, Eckert MA, Zhao W. To grab the stroma by the horns: from biology to cancer therapy with mesenchymal stem cells. Oncotarget 2013; 4(5): 651-64.
[http://dx.doi.org/10.18632/oncotarget.1040] [PMID: 23744479]

[122]   Zimmerlin L, Park TS, Zambidis ET, Donnenberg VS, Donnenberg AD. Mesenchymal stem cell secretome and regenerative therapy after cancer. Biochimie 2013; 95(12): 2235-45.
[http://dx.doi.org/10.1016/j.biochi.2013.05.010] [PMID: 23747841]

[123]   Larson BL, Ylostalo J, Lee RH, Gregory C, Prockop DJ. Sox11 is expressed in early progenitor human multipotent stromal cells and decreases with extensive expansion of the cells. Tissue Eng Part A 2010; 16(11): 3385-94.
[http://dx.doi.org/10.1089/ten.tea.2010.0085] [PMID: 20626275]

[124]   Gu Y, Li T, Ding Y, *et al.* Changes in mesenchymal stem cells following long-term culture *in vitro.* Mol Med Rep 2016; 13(6): 5207-15.
[PMID: 27108540]

[125]   Zhao Q, Gregory CA, Lee RH, *et al.* MSCs derived from iPSCs with a modified protocol are tumor-tropic but have much less potential to promote tumors than bone marrow MSCs. Proc Natl Acad Sci USA 2015; 112(2): 530-5.
[http://dx.doi.org/10.1073/pnas.1423008112] [PMID: 25548183]

[126]   Sánchez L, Gutierrez-Aranda I, Ligero G, *et al.* Enrichment of human ESC-derived multipotent mesenchymal stem cells with immunosuppressive and anti-inflammatory properties capable to protect against experimental inflammatory bowel disease. Stem Cells 2011; 29(2): 251-62.
[http://dx.doi.org/10.1002/stem.569] [PMID: 21732483]

[127]   Ludwig TE, Bergendahl V, Levenstein ME, Yu J, Probasco MD, Thomson JA. Feeder-independent culture of human embryonic stem cells. Nat Methods 2006; 3(8): 637-46.
[http://dx.doi.org/10.1038/nmeth902] [PMID: 16862139]

[128]   Olivier EN, Bouhassira EE. Differentiation of human embryonic stem cells into mesenchymal stem cells by the "raclure" method. Methods Mol Biol 2011; 690: 183-93.
[http://dx.doi.org/10.1007/978-1-60761-962-8_13] [PMID: 21042994]

[129]   Vasko T, Frobel J, Lubberich R, Goecke TW, Wagner W. iPSC-derived mesenchymal stromal cells are less supportive than primary MSCs for co-culture of hematopoietic progenitor cells. J Hematol Oncol 2016; 9: 43.
[http://dx.doi.org/10.1186/s13045-016-0273-2] [PMID: 27098268]

[130]   Fernández Vallone VB, Romaniuk MA, Choi H, Labovsky V, Otaegui J, Chasseing NA. Mesenchymal stem cells and their use in therapy: what has been achieved? Differentiation 2013; 85(1-2): 1-10.
[http://dx.doi.org/10.1016/j.diff.2012.08.004] [PMID: 23314286]

[131]   Frobel J, Hemeda H, Lenz M, *et al.* Epigenetic rejuvenation of mesenchymal stromal cells derived

from induced pluripotent stem cells. Stem Cell Reports 2014; 3(3): 414-22.
[http://dx.doi.org/10.1016/j.stemcr.2014.07.003] [PMID: 25241740]

[132]   Zhou T, Benda C, Duzinger S, *et al.* Generation of induced pluripotent stem cells from urine. J Am Soc Nephrol 2011; 22(7): 1221-8.
[http://dx.doi.org/10.1681/ASN.2011010106] [PMID: 21636641]

[133]   Wang L, Wang L, Huang W, *et al.* Generation of integration-free neural progenitor cells from cells in human urine. Nat Methods 2013; 10(1): 84-9.
[http://dx.doi.org/10.1038/nmeth.2283] [PMID: 23223155]

[134]   Lu J, Liu H, Huang CT, *et al.* Generation of integration-free and region-specific neural progenitors from primate fibroblasts. Cell Reports 2013; 3(5): 1580-91.
[http://dx.doi.org/10.1016/j.celrep.2013.04.004] [PMID: 23643533]

[135]   Amemori T, Ruzicka J, Romanyuk N, Jhanwar-Uniyal M, Sykova E, Jendelova P. Comparison of intraspinal and intrathecal implantation of induced pluripotent stem cell-derived neural precursors for the treatment of spinal cord injury in rats. Stem Cell Res Ther 2015; 6: 257.
[http://dx.doi.org/10.1186/s13287-015-0255-2] [PMID: 26696415]

[136]   Sareen D, Gowing G, Sahabian A, *et al.* Human induced pluripotent stem cells are a novel source of neural progenitor cells (iNPCs) that migrate and integrate in the rodent spinal cord. J Comp Neurol 2014; 522(12): 2707-28.
[http://dx.doi.org/10.1002/cne.23578] [PMID: 24610630]

[137]   All AH, Gharibani P, Gupta S, *et al.* Early intervention for spinal cord injury with human induced pluripotent stem cells oligodendrocyte progenitors. PLoS One 2015; 10(1): e0116933.
[http://dx.doi.org/10.1371/journal.pone.0116933] [PMID: 25635918]

[138]   Lee-Kubli CA, Lu P. Induced pluripotent stem cell-derived neural stem cell therapies for spinal cord injury. Neural Regen Res 2015; 10(1): 10-6.
[http://dx.doi.org/10.4103/1673-5374.150638] [PMID: 25788906]

[139]   López-Serrano C, Torres-Espín A, Hernández J, *et al.* Effects of the spinal cord injury environment on the differentiation capacity of human neural stem cells derived from induced pluripotent stem cells. Cell Transplant 2016; •••
[http://dx.doi.org/10.3727/096368916X691312] [PMID: 27075820]

[140]   Mauritz C, Schwanke K, Reppel M, *et al.* Generation of functional murine cardiac myocytes from induced pluripotent stem cells. Circulation 2008; 118(5): 507-17.
[http://dx.doi.org/10.1161/CIRCULATIONAHA.108.778795] [PMID: 18625890]

[141]   Narazaki G, Uosaki H, Teranishi M, *et al.* Directed and systematic differentiation of cardiovascular cells from mouse induced pluripotent stem cells. Circulation 2008; 118(5): 498-506.
[http://dx.doi.org/10.1161/CIRCULATIONAHA.108.769562] [PMID: 18625891]

[142]   Mauritz C, Schwanke K, Reppel M, *et al.* Generation of functional murine cardiac myocytes from induced pluripotent stem cells. Circulation 2008; 118(5): 507-17.
[http://dx.doi.org/10.1161/CIRCULATIONAHA.108.778795] [PMID: 18625890]

[143]   Dimos JT, Rodolfa KT, Niakan KK, *et al.* Induced pluripotent stem cells generated from patients with ALS can be differentiated into motor neurons. Science 2008; 321(5893): 1218-21.
[http://dx.doi.org/10.1126/science.1158799] [PMID: 18669821]

[144]   Yabe SG, Fukuda S, Takeda F, Nashiro K, Shimoda M, Okochi H. Efficient Generation of Functional Pancreatic β Cells from Human iPS Cells. J Diabetes 2017; 9(2): 168-79.
[http://dx.doi.org/10.1111/1753-0407.12400] [PMID: 27038181]

[145]   Sauer V, Roy-Chowdhury N, Guha C, Roy-Chowdhury J. Induced pluripotent stem cells as a source of hepatocytes. Curr Pathobiol Rep 2014; 2(1): 11-20.
[http://dx.doi.org/10.1007/s40139-013-0039-2] [PMID: 25650171]

[146]   Ao Y, Mich-Basso JD, Lin B, Yang L. High efficient differentiation of functional hepatocytes from

porcine induced pluripotent stem cells. PLoS One 2014; 9(6): e100417.
[http://dx.doi.org/10.1371/journal.pone.0100417] [PMID: 24949734]

[147] Sancho-Bru P, Roelandt P, Narain N, *et al.* Directed differentiation of murine-induced pluripotent stem cells to functional hepatocyte-like cells. J Hepatol 2011; 54(1): 98-107.
[http://dx.doi.org/10.1016/j.jhep.2010.06.014] [PMID: 20933294]

[148] Nakamori D, Takayama K, Nagamoto Y, *et al.* Hepatic maturation of human iPS cell-derived hepatocyte-like cells by ATF5, c/EBPα, and PROX1 transduction. Biochem Biophys Res Commun 2016; 469(3): 424-9.
[http://dx.doi.org/10.1016/j.bbrc.2015.12.007] [PMID: 26679606]

[149] Rosen JM, Jordan CT. The increasing complexity of the cancer stem cell paradigm. Science 2009; 324(5935): 1670-3.
[http://dx.doi.org/10.1126/science.1171837] [PMID: 19556499]

[150] Reya T, Morrison SJ, Clarke MF, Weissman IL. Stem cells, cancer, and cancer stem cells. Nature 2001; 414(6859): 105-11.
[http://dx.doi.org/10.1038/35102167] [PMID: 11689955]

[151] Sales KM, Winslet MC, Seifalian AM. Stem cells and cancer: an overview. Stem Cell Rev 2007; 3(4): 249-55.
[http://dx.doi.org/10.1007/s12015-007-9002-0] [PMID: 17955391]

[152] Tilkorn DJ, Lokmic Z, Chaffer CL, Mitchell GM, Morrison WA, Thompson EW. Disparate companions: tissue engineering meets cancer research. Cells Tissues Organs (Print) 2010; 192(3): 141-57.
[http://dx.doi.org/10.1159/000308892] [PMID: 20357428]

[153] Salter E, Goh B, Hung B, Hutton D, Ghone N, Grayson WL. Bone tissue engineering bioreactors: a role in the clinic? Tissue Eng Part B Rev 2012; 18(1): 62-75.
[http://dx.doi.org/10.1089/ten.teb.2011.0209] [PMID: 21902622]

[154] Petrovic V, Zivkovic P, Petrovic D, Stefanovic V. Craniofacial bone tissue engineering. Oral Surg Oral Med Oral Pathol Oral Radiol 2012; 114(3): e1-9.
[http://dx.doi.org/10.1016/j.oooo.2012.02.030] [PMID: 22862985]

[155] Xu W, Hu X, Pan W. Tissue engineering concept in the research of the tumor biology. Technol Cancer Res Treat 2014; 13(2): 149-59.
[http://dx.doi.org/10.7785/tcrt.2012.500363] [PMID: 23862747]

[156] Horch RE, Boos AM, Quan Y, *et al.* Cancer research by means of tissue engineering--is there a rationale? J Cell Mol Med 2013; 17(10): 1197-206.
[http://dx.doi.org/10.1111/jcmm.12130] [PMID: 24118692]

[157] Kao TC, Lee HH, Higuchi A, *et al.* Suppression of cancer-initiating cells and selection of adipose-derived stem cells cultured on biomaterials having specific nanosegments. J Biomed Mater Res B Appl Biomater 2014; 102(3): 463-76.
[http://dx.doi.org/10.1002/jbm.b.33024] [PMID: 24039170]

[158] Shigdar S, Li Y, Bhattacharya S, *et al.* Inflammation and cancer stem cells. Cancer Lett 2014; 345(2): 271-8.
[http://dx.doi.org/10.1016/j.canlet.2013.07.031] [PMID: 23941828]

[159] Medema JP, Vermeulen L. Microenvironmental regulation of stem cells in intestinal homeostasis and cancer. Nature 2011; 474(7351): 318-26.
[http://dx.doi.org/10.1038/nature10212] [PMID: 21677748]

[160] Beachy PA, Karhadkar SS, Berman DM. Tissue repair and stem cell renewal in carcinogenesis. Nature 2004; 432(7015): 324-31.
[http://dx.doi.org/10.1038/nature03100] [PMID: 15549094]

[161] Yan Y, Zuo X, Wei D. Concise Review: Emerging Role of CD44 in Cancer Stem Cells: A Promising

Biomarker and Therapeutic Target. Stem Cells Transl Med 2015; 4(9): 1033-43.
[http://dx.doi.org/10.5966/sctm.2015-0048] [PMID: 26136504]

[162]   Yang X, Sarvestani SK, Moeinzadeh S, He X, Jabbari E. Three-dimensional-engineered matrix to study cancer stem cells and tumorsphere formation: effect of matrix modulus. Tissue Eng Part A 2013; 19(5-6): 669-84.
[http://dx.doi.org/10.1089/ten.tea.2012.0333] [PMID: 23013450]

[163]   Gao P, Jiang D, Liu W, Li H, Li Z. Urine-derived Stem Cells, A New Source of Seed Cells for Tissue Engineering. Curr Stem Cell Res Ther 2016; 11(7): 547-3.
[http://dx.doi.org/10.2174/1574888X10666150220161506] [PMID: 25697496]

[164]   Takakura N. Formation and regulation of the cancer stem cell niche. Cancer Sci 2012; 103(7): 1177-81.
[http://dx.doi.org/10.1111/j.1349-7006.2012.02270.x] [PMID: 22416970]

[165]   Chhabra R, Saini N. MicroRNAs in cancer stem cells: current status and future directions. Tumour Biol 2014; 35(9): 8395-405.
[http://dx.doi.org/10.1007/s13277-014-2264-7] [PMID: 24964962]

[166]   Islam F, Gopalan V, Smith RA, Lam AK. Translational potential of cancer stem cells: A review of the detection of cancer stem cells and their roles in cancer recurrence and cancer treatment. Exp Cell Res 2015; 335(1): 135-47.
[http://dx.doi.org/10.1016/j.yexcr.2015.04.018] [PMID: 25967525]

[167]   Loessner D, Holzapfel BM, Clements JA. Engineered microenvironments provide new insights into ovarian and prostate cancer progression and drug responses. Adv Drug Deliv Rev 2014; 79-80: 193-213.
[http://dx.doi.org/10.1016/j.addr.2014.06.001] [PMID: 24969478]

[168]   Guarino V, Cirillo V, Altobelli R, Ambrosio L. Polymer-based platforms by electric field-assisted techniques for tissue engineering and cancer therapy. Expert Rev Med Devices 2015; 12(1): 113-29.
[http://dx.doi.org/10.1586/17434440.2014.953058] [PMID: 25487005]

[169]   da Rocha EL, Porto LM, Rambo CR. Nanotechnology meets 3D *in vitro* models: tissue engineered tumors and cancer therapies. Mater Sci Eng C 2014; 34: 270-9.
[http://dx.doi.org/10.1016/j.msec.2013.09.019] [PMID: 24268259]

[170]   Burdett E, Kasper FK, Mikos AG, Ludwig JA. Engineering tumors: a tissue engineering perspective in cancer biology. Tissue Eng Part B Rev 2010; 16(3): 351-9.
[http://dx.doi.org/10.1089/ten.teb.2009.0676] [PMID: 20092396]

[171]   Martínez-Ramos C, Lebourg M. Three-dimensional constructs using hyaluronan cell carrier as a tool for the study of cancer stem cells. J Biomed Mater Res B Appl Biomater 2015; 103(6): 1249-57.
[http://dx.doi.org/10.1002/jbm.b.33304] [PMID: 25350680]

[172]   Xu X, Farach-Carson MC, Jia X. Three-dimensional *in vitro* tumor models for cancer research and drug evaluation. Biotechnol Adv 2014; 32(7): 1256-68.
[http://dx.doi.org/10.1016/j.biotechadv.2014.07.009] [PMID: 25116894]

[173]   Cheng Y, Zheng F, Lu J, *et al.* Bioinspired multicompartmental microfibers from microfluidics. Adv Mater 2014; 26(30): 5184-90.
[http://dx.doi.org/10.1002/adma.201400798] [PMID: 24934291]

[174]   Kim D, Wu X, Young AT, Haynes CL. Microfluidics-based *in vivo* mimetic systems for the study of cellular biology. Acc Chem Res 2014; 47(4): 1165-73.
[http://dx.doi.org/10.1021/ar4002608] [PMID: 24555566]

[175]   Chung BG, Lee KH, Khademhosseini A, Lee SH. Microfluidic fabrication of microengineered hydrogels and their application in tissue engineering. Lab Chip 2012; 12(1): 45-59.
[http://dx.doi.org/10.1039/C1LC20859D] [PMID: 22105780]

[176]   Zhao Y, Cheng Y, Shang L, Wang J, Xie Z, Gu Z. Microfluidic synthesis of barcode particles for

multiplex assays. Small 2015; 11(2): 151-74.
[http://dx.doi.org/10.1002/smll.201401600] [PMID: 25331055]

[177] Inamdar NK, Borenstein JT. Microfluidic cell culture models for tissue engineering. Curr Opin Biotechnol 2011; 22(5): 681-9.
[http://dx.doi.org/10.1016/j.copbio.2011.05.512] [PMID: 21723720]

[178] Lee MG, Shin JH, Bae CY, Choi S, Park JK. Label-free cancer cell separation from human whole blood using inertial microfluidics at low shear stress. Anal Chem 2013; 85(13): 6213-8.
[http://dx.doi.org/10.1021/ac4006149] [PMID: 23724953]

[179] Zhang Z, Nagrath S. Microfluidics and cancer: are we there yet? Biomed Microdevices 2013; 15(4): 595-609.
[http://dx.doi.org/10.1007/s10544-012-9734-8] [PMID: 23358873]

[180] Villasante A, Vunjak-Novakovic G. Tissue-engineered models of human tumors for cancer research. Expert Opin Drug Discov 2015; 10(3): 257-68.
[http://dx.doi.org/10.1517/17460441.2015.1009442] [PMID: 25662589]

[181] Meacham CE, Morrison SJ. Tumour heterogeneity and cancer cell plasticity. Nature 2013; 501(7467): 328-37.
[http://dx.doi.org/10.1038/nature12624] [PMID: 24048065]

[182] Sanna V, Pala N, Sechi M. Targeted therapy using nanotechnology: focus on cancer. Int J Nanomedicine 2014; 9: 467-83.
[PMID: 24531078]

[183] Grodzinski P, Farrell D. Future opportunities in cancer nanotechnology--NCI strategic workshop report. Cancer Res 2014; 74(5): 1307-10.
[http://dx.doi.org/10.1158/0008-5472.CAN-13-2787] [PMID: 24413533]

[184] Menter DG, Patterson SL, Logsdon CD, Kopetz S, Sood AK, Hawk ET. Convergence of nanotechnology and cancer prevention: are we there yet? Cancer Prev Res (Phila) 2014; 7(10): 973-92.
[http://dx.doi.org/10.1158/1940-6207.CAPR-14-0079] [PMID: 25060262]

CHAPTER 2

Dental Tissue Engineering and Regeneration; Perspectives on Stem Cells, Bioregulators, and Porous Scaffolds

Perihan Selcan Gungor-Ozkerim[1] and Abdulmonem Alshihri[2,3,*]

[1] *Harvard-MIT Division of Health Sciences and Technology, Massachusetts Institute of Technology, Cambridge, MA, 02139, USA*

[2] *Department of Restorative and Biomaterials, Harvard School of Dental Medicine, Boston, MA, 02115 USA*

[3] *Department of Prosthetic and Biomaterial Sciences, College of Dentistry, King Saud University, Riyadh 11545, Saudi Arabia*

Abstract: Dental/orofacial tissue engineering is an emerging field that offers alternative solutions for dental problems resulting from pathologies such as caries, trauma, periodontal disease and others. Various stem cell types such as bone marrow stem cells (BMSCs) and adipose tissue-derived stem cells (ADSCs) can be employed as cell sources for dental tissue repair. In particular, dental tissue-derived stem cells such as dental pulp stem cells (DPSCS) and dental follicle stem cells (DFSCs) can be utilized due to their favorable origin and properties. On the other hand, natural and synthetic polymers are used to fabricate 3D dental tissue scaffolds to support cellular activities. Several bioregulators such as cytokines and growth factors can also be incorporated to induce cells interaction and cell-scaffold integration. The literature on the regeneration of dental tissues via tissue engineering principles presents numerous results that are superior to the traditional methods, which are vital for advancing dental therapy. Herein, the types, properties, and applications of dental stem cells (DSCs) were reviewed, as well as the tissue engineering and regeneration strategies for different types of dental tissues.

Keywords: 3D scaffolds, Alveolar bone, Dental pulp, Dental tissue engineering, Dental follicle, Growth factors, Intrabony defect, Periodontium, Periodontal ligament, Stem cells, Tissue regeneration.

---

* **Corresponding author Abdulmonem A. Alshihri:** Restorative and Biomaterial Sciences, Harvard School of Dental Medicine, Boston, MA, 02115, USA, Department of Prosthetic and Biomaterial Sciences, College of Dentistry, King Saud University, Riyadh, 11545, Saudi Arabia; Tel: +966114677333; Fax: +966114679015; E-mail: Monem.alshihri@post.harvard.edu

Mehdi Razavi (Ed.)
All rights reserved-© 2017 Bentham Science Publishers

INTRODUCTION

Tissue engineering is an emerging field that utilizes cells, scaffolds, and bioactive signaling factors - individually or combined- for tissues replacement, repair and regenerative purposes. Dental tissue engineering and regenerative dental medicine represent translational research and interventional therapeutic concepts to overcome several clinical limitations. These engineering and regenerative concepts cover a wide range of applicability in different dentoalveolar and oro-facial branches of dentistry [1 - 3]. The numerous applications of dental and mesenchymal stem cells are attributed to their common properties of self-renewal, high proliferation rate, and multilineage differentiation ability. Therefore, there is a considerable interest in their potentials as an alternative stem cell source to tackle issues involving mesenchymal or non-mesenchymal cell derivatives [4, 5]. Dental stem cells can be isolated from different dental tissues, such as dental pulp, periodontal ligament or tooth follicle.

Different clinical situations could result in teeth removal in which these teeth could serve as a source of stem cells. For example extracted third molars (wisdom teeth), teeth that are extracted for orthodontic reasons, and exfoliated deciduous teeth. Isolation and cryopreservation of these cells are of researchers' interest because of their relative ease of collection, attractive applications and reparative potential. In this chapter, various dental tissue-derived stem cells, their applications and regenerative capacity in tissue engineering are presented.

DENTAL STEM CELLS

Stem cells, in general, possess main features of self-renewal and multilineage-differentiation capability. Stem cells have been isolated and differentiated from different parts of the body. Hence, they are categorized based on their source and location. They however differ according to lineages of differentiation [6]. Various populations of stem cells have been isolated and grown from different parts of dental tissues and are classified accordingly as; dental pulp stem cells (DPSCs), periodontal ligament stem cells (PDLSCs), dental follicle stem cells (DFSCs), stem cells from human exfoliated deciduous teeth (SHED), stem cells from apical papilla (SCAP) and (BMSCs) bone marrow. Moreover, dental stem cells have attracted a considerable interest in their use for other, non-dental, medical applications as well [4].

Dental Pulp Stem Cells

The dental pulp is the unmineralized connective tissue occupying the central core of a tooth. It is highly vascularized and innervated tissue of ectomesenchymal origin. The dental pulp provides sensory and protective functions, and is responsible for the formation and nutrition of the surrounding dentin [7, 8]. Histologically, dental pulp tissue is composed of various types of cells. Fibroblasts and undifferentiated mesenchymal cells are the predominant cells of the pulp, which correspond to its reparatory and defensive functions [9]. These mesenchymal progenitor cells are referred to as dental pulp stem cells (DPSCs). In 2000, DPSCs were discovered and studied by Gronthos *et al* [10]. Their findings revealed that DPSCs are clonogenic with high proliferation capacity, and are able to regenerate new tissue. Numerous studies [11 - 14] demonstrated the potential of DPSCs as qualified stem cells because of their multi-lineage differentiation ability and self-renewal capacity. Moreover, DPSCs exhibited a significantly higher proliferation rate and better multilineage differentiation compared to bone marrow stem cells (BMSCs) [11, 14, 15]. These qualities were related to the neural crest origin and embryonic cell-like behavior of DPSCs, as opposed to the BMSCs' mesodermal origin [12, 16, 17]. Another important aspect of DPSCs is their abundancy, which is a critical factor for cellular pool availability. The availability of DPSCs is of importance for conducting and translating research into clinical therapeutic applications [18]. However, the clinical use of DPSCs could be associated with potential immunosuppressive [19] and immunomodulating [20] responses, which are crucial limitations.

Similar to DPSCs isolated from human permanent teeth [21], stem cells can also be derived from human deciduous (primary) teeth pulp. These cells are called stem cells from human exfoliated deciduous teeth (SHED) [7, 18, 22]. In 2003, Dr. Shi discovered SHED. These cells, compared to DPSCs, possess higher proliferation and less maturation rates during cell passage, as well as higher potentials for differentiation to a variety of cells [23].

Another form of tooth-related stem cells were found in apical papilla (SCAP) by Sonoyama *et al.* These cells are able to differentiate into mineralized tissue-producing cells where they reside in teeth's roots to aid in root formation [8, 24, 25]. Furthermore, a group of researchers reported a discovery of embryonic-like stem cells in human dental pulp, which were characterized by embryonic lineage markers [26].

Although DPSCs are relatively new compared to their counterparts, such as BMSCs or cord blood stem cells, they are employed in a wide range of applications in tissue engineering. DPSCs including SHED and SCAP cells were particularly utilized to obtain osteoblastic lineage directed cells for bone [27 - 30] and dentin regeneration [31, 32]. Another attractive field for employing DPSCs is neural regeneration for their neural crest origin [11, 33 - 35]. DPSCs were also established for chondrogenic [17, 21, 36], myocytic [12, 17, 21], hepatocytic [12, 13, 37], adipogenic [11, 13] lineages differentiation, and recently epidermal differentiation [38]. Thus, they can be utilized for treating bone, cartilage, skin, muscle, liver defects, neurodegenerative disorders, and also can be used for fat tissue engineering. Some other specific applications of DPSCs are therapies for ocular [39, 40], lung [41], kidney [42], diabetes [43], immune disorders [20], cardiac injuries [44], urinary bladder injuries [45], and autism spectrum disorders [46]. Additionally, from dentinogenesis and dental tissue regeneration perspective, numerous studies have reported dentin and pulpal tissue regeneration with DPSCs [10, 11, 47 - 50]. Yu *et al.* investigated odontoblastic lineage differentiation of DPSCs via tooth germ cells conditioned-media [47]. They obtained odontogenic expressions *in vitro* as well as dentin-pulp complex development *in vivo*. They also compared the odontogenic capability of DPSCs and BMSCs in the presence of SCAPs, they demonstrated the superiority of DPSCs for such purposes [48]. Batouli *et al.* reported that the transplanted DPSCs expressed high dentin sialoprotein *in vivo*, and observed the formation of a dentin-like layer [49]. In another study, DPSCs and SCAPs were used in a mouse model, where it resulted in a *de novo* regeneration of a pulp-like structure in the emptied root canal [50]. Furthermore, a uniform dentin-like structure was formed on the canal walls, which is attributed to the physiological function of the pulp. The formation of aligned and bilayered tubular networks that mimicked the native tubular dentine was also established [51]. In another study, researchers treated DPSCs with calcium hydroxide, a conditional liner for deep dental cavities, where an increase in cell proliferation and differentiation was found [52].

Because of the various applications and reparative potentials of DPSCs, cryopreservation of these cells attracts the attention of researchers from different fields. Specific characteristics of DPSCs can be retained even after long-term cryopreservation [8], making them a strong candidate for cryopreserved cell banks for adult tissue regeneration [18]. Particularly, preservation of SCAPs isolated from third-molars can be a valuable source for autologous future therapy [24].

Periodontal Ligament Stem Cells

The periodontium refers to the tissue zone which supports and surrounds teeth; providing its stability in the alveolar bone. It consists of four different parts; gingiva, alveolar bone, periodontal ligaments (PDLs) and cementum. PDLs are a major component of the periodontium. They are highly neurovascularized connective tissue with rich cellular and collagenous elements. PDLs are located between a tooth's root and alveolar bone, which provide a supportive attachment to the surrounding bone [53 - 56]. Similar to dental pulp, PDLs are of a neural crest origin [57]. PDL has a heterogeneous histology composed of fibroblastic cells, immunogenic cells and, undifferentiated mesenchymal cells, such as PDLSCs. These stem cells are the resource for cementoblasts and osteoblasts, which maintain and repair the periodontium [55, 58]. These PDL cells have a higher metabolic activity than cells in other tissues; for instance, their collagen turnover is 15 times more than that of skin's fibroblasts [59]. PDLSCs were first reported by Seo *et al.* [60] a few years after DPSCs were discovered. Their reports and following studies revealed that PDLSCs are similar to DPSCs and BMSCs in terms of expressing mesenchymal stem cell-specific markers, and having high proliferation and multi-lineage differentiation capability [54, 60 - 62]. In addition, PDLSCs possess immunomodulatory behavior [63] and offer an easily accessible source via noninvasive or minimal invasive retrieval [54, 60]. In contrast to abundance of DPSCs, PDLSCs exhibit a relatively lower cell number available for clinical use [53, 55]. Furthermore, the proliferation and regeneration ability of PDLSCs decrease as the donor's age increases [64].

PDLSCs were successfully differentiated into osteogenic [60, 65], chondrogenic [62, 66], adipogenic [53, 67], neurogenic [68], hepatic [69], vascular [70], and pancreatic [71] lineages under suitable conditions. Although the proliferation rates and clonogenic ability of PDLSCs were significantly higher than BMSCs, their *in vitro* osteogenic capacity was lower than that of BMSCs [63, 72, 73]. When compared to DPSCs, both proliferation and osteogenic differentiation abilities of PDLSCs were lower than that of DPSCs [74]. On the other hand, it was reported that some *in vitro* techniques such as co-culturing PDLSCs with endothelial cells [75, 76], or exposing them to hypoxic conditions [76, 77], significantly improved the osteogenic differentiation of PDLSCs. Some bioactive factors such as basic fibroblast growth factor (bFGF) promoted the osteogenic maturation of PDLSCs [78]. Furthermore, PDLSCs were recently utilized in *in vitro* and *in vivo* for tendon regeneration, and displayed a significantly higher performance than BMSCs in the same microenvironment [79].

One of the most promising applications of PDLSCs is the regeneration of the periodontal tissue components [53]. Destruction of the periodontium can result from dental diseases such as periodontitis, which can lead to tooth loss and other local and systemic concerns in later stages [54, 58]. Currently, there is no conventional therapy for an efficient and complete PDL regeneration [80]. Thus, PDLSCs can provide a valuable therapeutic tool for reconstruction of these tissues [60]. Liu *et al.* utilized PDLSCs to treat periodontal defects in a porcine model [81]. PDLSCs were derived from an autologous extracted tooth, proliferated *ex vivo* and then transferred to the defect zone. This approach has resulted in some periodontal tissue regeneration cues. Another study demonstrated cementum-like tissue formation in a mouse model after PDLSCs transplantation in the presence of hydroxyapatite (HAp)/beta-tricalcium phosphate (β-TCP) composite [82]. Sonoyama *et al.* used PDLSCs with SCAPs to obtain a root periodontal complex. They suggested that the hybrid cell system improved tissue regeneration *in vivo*

[83]. Similarly, another study proposed a multilayer cell pellet construct in which each layer contains PDLSCs of microenviroment to form a 3D cement-PDL complex [84]. A different study used cell sheets instead of cell pellets to obtain the multilayered complex structure. The different layers differentiated into different tissues of PDL and cementum [85, 86]. Similarly, PDLSCs sheet method was applied in a dog model for autologous periodontal therapy, it resulted in periodontal regeneration (PDL and cementum formation with bone healing) [87]. Furthermore, alveolar bone regeneration was established by subcutaneous transplantation of rat molars. It showed a differentiation of PDLSCs on the extracted tooth surface, migrating to the adjacent alveolar bone [88].

Similar to DPSCs, cryopreservation of PDLSCs is of interest to researchers due to their regeneration potential. It was reported that characteristic properties of cryopreserved PDLSCs were maintained successfully. And they can be saved for either future autogenic [89, 90] or allogeneic [91] clinical applications.

Dental Follicle Stem Cells

The dental follicle (DF) is a sac-like loose connective tissue of ectomesenchymal cells and fibers present in a tooth bud. It is located within the alveolar bone surrounding the enamel organ and dental papilla of a developing tooth [92, 93]. The DF gives rise to root cementum, periodontal ligament and the alveolar bone surrounding it. It can be isolated from third-molars that are commonly extracted in dental practice. In such scenarios, the DF can be classified into two different regions; as a periapical part, which is in proximity to the apical part of developing tooth root and a coronal part, which is in proximity to the developing periodontium [94]. It was reported that the coronal part of the DF plays a significant role in tooth eruption process. On the other hand, the periapical part stimulates bone growth and root development [94 - 96]. Similar to dental pulp and periodontal ligament, the DF harbors undifferentiated multipotent mesenchymal stem cells (MSCs). These MSCs originate from the neural crest and are called dental follicle stem cells (DFSCs) [94]. As a subgroup of DFSCs, the stem cells derived from the apical end of the DF during its development are called periapical follicle stem cells (PAFSCs) [97]. DFSCs exhibited a higher *in vitro* proliferation rate than that of DPSCs [98 - 100] and PDLSCs [99]. However, DFSCs demonstrated immunomodulatory behavior similar to other mesenchymal stem cells [101]. Additionally, DFSCs showed an immunomodulatory effect that is higher than that of DPSCs and SHEDs [102]. Hence, they can be beneficial for the therapy of autoimmune, allergic, inflammatory, and other related diseases. Both *in vitro* and *in vivo* differentiations of DFSCs were reported for several lineages including osteogenic [93, 99, 100, 103], neurogenic [104 - 107], adipogenic [99, 100, 102, 107], and chondrogenic [100, 102, 108] cells. Park *et al.*

reported a superior osteocalcin and calcium content associated with DFSCs when compared with the osteogenesis capacity of other MSCs isolated from skin and bone marrow *in vivo* [108]. Such findings revealed a remarkable bone tissue engineering potential of DFSCs [109]. Similarly, DFSCs were incorporated in poly-epsilon-caprolactone (PCL) scaffolds, and this complex was successfully used for craniofacial bone therapy in an animal model [110].

DFSCs and PAFSCs demonstrate a great differentiation capacity into various types of dental tissue cells, which make them superior options for dental tissue regeneration [94]. Under suitable stimulatory conditions, it was reported that DFSCs produced dentin-like tissues *in vivo*, which proves their dentin regeneration potential [111, 112]. Yokoi *et al.* isolated DFSCs from tooth germs of a mouse incisor, and found a PDL progenitor-specific gene expression and PDL-like tissues formation [113]. By using cell sheets method, DFSCs were co-cultured with epithelial root sheath cells and transferred into a rat model. It resulted in cementum and PDL-like tissues formation *in vivo* [114]. Similarly, PAFSCs produced cementum and PDL-like tissues *in vivo* [97]. Another study utilized growth factors to differentiate DFSCs into periodontal tissue-related lineages [115]. In addition, DFSCs combined with inductive dentin matrix scaffolds were used for root regeneration, which can be of value for tooth loss conditions [112, 116, 117]. Considering all these positive results, DFSCs are another favorable cell source for tissue engineering. Unerupted third molars or deciduous teeth, which are generally extracted and discarded as a medical waste, can be utilized as an easily accessible source of DFSCs and PAFSCs for various clinical applications [98]. It was demonstrated that cryopreserved and native DFSCs display similar growth, immunologic, and differentiation properties [101]. Thus, DFSCs are candidates for cell banking where they can be stocked for autogenic and allogeneic therapy needs.

Orofacial Bone Marrow Stem Cells

Mesenchymal stem cells are named for their mesodermal origin and their differentiation into mesodermal cell types. They can be found in almost all adult tissues. Due to their multilineage differentiation ability, including meso-, ecto- and endodermal lineages, they are used pre-clinical and clinical studies for various purposes [118, 119]. MSCs were first isolated, identified and characterized from bone marrow (BMSCs). They have been extensively utilized and studied in tissue engineering principles and applications as BMSCs [120]. Similar to long bones, jawbones or orofacial bones (maxilla, mandible and alveolar bone) can also be a source of BMSCs [121, 122]. Orofacial BMSCs originate from the neural crest mesenchyme similar to the dental stem cells. On the other hand, BMSCs derived from other bones are originating from mesoderm

[123]. In addition, it was reported that maxillomandibular-derived BMSCs possess higher proliferation rates, higher expression of several major osteogenic markers (alkaline phosphatase and osteopontin), and more delayed senescence than those of axial bones (iliac crest) [124, 125]. Similarly, alveolar bone-derived BMSCs have distinct gene expression levels and differentiation capacity compared to long bones. It is mainly due to their different skeletal locations and hence, different environmental influences such as mechanical load and developmental origin [126 - 128]. It was shown that the stimulatory effects of bone morphogenetic protein (BMP) on orofacial BMSCs is significantly higher than that of iliac BMSCs. This implies that orofacial BMSCs require a lower level of *in vitro* induction, which is reflective of their osteogenic capacity [125, 129]. The same group of researchers further presented that the orofacial BMSCs had similar titanium attachment profiles as the iliac BMSCs. Also, the osteogenic potential of titanium-attached orofacial BMSCs was greater than that of iliac BMSCs [130]. These findings suggest that orofacial BMSCs can be used successfully in titanium implant-based tissue engineering. Moreover, other studies reported that orofacial BMSCs possess more pronounced immunosuppressive behavior than that of iliac BMSCs in a murine model [131]. It was also reported that orofacial BMSCs can be genetically modified in order to improve bone regeneration. This was achieved for autologous therapy of a mandibular defect in a mouse model [132]. In contrast to their strong osteogenic and bone regeneration potential [131, 133], orofacial BMSCs exhibited lower chondrogenic and adipogenic differentiation ability compared to iliac BMSCs [126, 134]. Lower adipogenesis may be a favorable property preventing unwanted fat tissue development during regeneration process in the BMSCs transplantation area [134]. Another study compared orofacial and iliac BMSCs for their irradiation response. Results revealed that orofacial BMSCs have a better resistance to radiation with faster recovery compared to iliac BMSCs [135]. Aspiration of dental alveolar BMSCs is less invasive and less painful compared to that of iliac BMSCs. This is an important advantage in clinical application and patient morbidity. Moreover, alveolar BMSCs can be isolated from patients during molar extraction and cryopreserved for future clinical use [134]. Similarly, maxillary and mandibular BMSCs can be reached during relevant dental procedures. Although axial and appendicular BMSCs are commonly used in various clinical applications, orofacial bones may also be utilized as an alternative source of BMSCs. Particularly, for bone tissue engineering, dental tissue engineering, and orofacial reconstructive therapies [121, 132].

In conclusion, all types of DSCs have favorable properties for tissue engineering and regenerative medicine applications. They have main stem cell properties and are suitable for minimally invasive cell banking.

DENTAL PULP TISSUE ENGINEERING AND REGENERATION

As mentioned in the Dental Pulp Stem Cells section, dental pulp is the unmineralized neurovascular connective tissue located in the core zone of a tooth. It maintains the vitality and functionality of a tooth via its vascular network content within the elastic and collagen fiber structures. In addition, the nerve network maintains its reflex to mechanical, thermal, or chemical signals and governs its stimulus-response mechanism [136]. Although the dental pulp is surrounded and protected by mineralized tissues of dentin and enamel, it is under the risk of bacterial invasion (dental caries, trauma, *etc.*). These various external factors can cause different stages of pulpal inflammation known as pulpal infection or pulpitis [137]. The dental pulp tissue contains blood vessels and immune defense components. When the body's immune system cannot clear the whole infection region, necrosis of the pulp occurs [137, 138]. The current endodontic treatment method is called pulpectomy, also known as root canal therapy. It is the instrumental removal of the contaminated pulp tissue and replaced by filling the void area with a bioinert nonliving material, losing the tissue irreversibly [138, 139]. Because the dental pulp has an integral function for tooth viability, regeneration of this tissue is one of the main goals of dental tissue engineering [136]. The size of the dental pulp tissue is very small with respect to other larger tissues. This may be advantageous for regeneration via tissue engineering approaches. The aim of regeneration is revascularization and recellularization of dental pulp tissue, which would be superior to current therapy [137, 139]. Following the general principles of tissue engineering, stem cell scaffolds and bioregulators are utilized in tissue engineering applications either alone or combined. Dental pulp regeneration approach can also benefit from these strategies.

Stem Cells for Dental Pulp Tissue Engineering

Stem cells are the most frequently referenced cell source for tissue engineering applications. It is due to their undifferentiated profile with a multilineage differentiation ability and self-renewal capacity [139]. A modality in current endodontic therapy is which dental pulp is regenerated via grafting MSCs into the cleaned root canals [140]. Several studies reported that MSCs, including DPCSs, PDLSCs and DFSCs, could be found in specific sites of a tooth. DPCSs originate from dental pulp tissue and have the ability to differentiate into odontoblasts. DPCSs (including SHED and SCAP) seem to be the stem cell of choice for dental pulp regeneration [138]. In addition, DPCSs possess angiogenic and neurogenic potentials [140]. They are the most studied and well-characterized class of DSCs. Several studies investigated dentin and pulp tissue regeneration potential of DPSCs [10, 11, 47 - 49, 52]. There are also several *in vivo* studies in animal models that utilized DPSCs for pulp tissue regeneration. Nakashima *et al.* reported that transplanted autologous DPSCs stimulated pulp tissue regeneration in a dog's root canal space following pulpectomy. In addition, they reported similar *in vitro* performances of DPSCs from dogs of different ages. However, *in vivo* regeneration success was decreased in the older dogs [141]. Similarly, another group tested regeneration performance of autologous DPSCs in a canine model [142]. They isolated DPSCs from first molars and cultured them to increase the cell numbers. Cells were then transplanted into the pulp-free canal within a gel foam scaffold. This resulted in a pulp-like structure including dentin-like tissue and blood vessels. Another group utilized DPSCs and SCAPs in a mouse-model for *de novo* pulp regeneration of an empty root canal. They obtained a uniform dentin-like structure and vascularized pulp like structure [50]. Furthermore, Yu *et al.* demonstrated odontogenic capacity of autologous DPSCs in a rat model by transplanting them into renal capsules [47, 48]. In another study, DPSCs isolated from human third molars were expanded and implanted subcutaneously along with HAp - β-TCP particles in immunocompromised mice. Results showed a donor derived pulp-like structures with clear odontoblast layers [82]. Similarly, SHED seeded on scaffolds that were transplanted into immunodeficient mice resulted in odontoblast-like cells [82]. Although DPSCs-based tissue engineering of dental pulp seems like an ideal approach, the autologous availability of human DPSCs are limited as discussed previously [139]. Because of that limitation, another research group compared DPSCs derived from normal and inflamed pulps with respect to their *in vitro* and *in vivo* capacities [143]. Although some of the *in vitro* stem cell properties were slightly less in the inflamed pulp cells, they maintained tissue regeneration capabilities; making them an alternative candidate as a source of DPSCs. BMSCs were also reported as another stem cell source for dental pulp regeneration. Xu *et al.* demonstrated that systemically transplanted-BMSCs induced pulp regeneration in a mouse model after pulpectomy [144]. The labeled BMSCs were identified by a fluorescence microscope. In another study, the same group used a similar methodology, cell homing, in a mouse model. They reported that stromal cell derived factor-1 (SDF-1) improved BMSCs migration to the pulpal system and formation of blood vessels with better pulp regeneration [145]. Another study utilized a canine model, pulp regeneration from BMSCs was compared with DPSCs and adipose tissue derived stem cells (ADSCs). It was concluded that DPSCs and ADSCs were superior to BMSCs [146]. Furthermore, Ravindran *et al.* used DPSCs and PDLSCs together in a blended collagen/chitosan scaffold for dental pulp tissue engineering and achieved promising results [138].

The studies of pulp tissue regeneration via stem cells in human are currently in clinical trial. For utilizing this treatment modality, large-scale clinical results are needed prior to the approval of the Food and Drug Administration (FDA). On the other hand, banking of autologous stem cells including cryopreserved DPSCs has been established where individuals could keep their own stem cells for future clinical needs [139].

Scaffolds for Dental Pulp Tissue Engineering

As being one of the main components of tissue engineering, scaffolds provide the required biological, chemical, and spatial constructive support for cells. Scaffolds provide a framework of tri-dimensional (3D) microenvironment for cells to establish an extracellular matrix (ECM) [111]. The minimum requirements for scaffold biomaterials to be successful are being biocompatible, biodegradable, porous, and inductive to regeneration of target tissue. Further specifications can be triggered by the biomaterial selection. In addition to biomaterial selection, different fabrication techniques such as electrospinning or solvent-casting can be utilized according to cells specification. Tissue engineering scaffold materials can be natural or synthetic-made. A blend of these materials can be considered and preferred as hybrid scaffold materials [137].

Due to its importance, abundance, and ease production, collagen is one of the mostly used natural proteins in tissue engineering. It is utilized for dental pulp tissue regeneration as well. Prescott *et al.* combined a collagen-based scaffold with the growth factor dentin matrix protein 1 (DMP1) and DPSCs to promote pulp tissue regeneration [147]. A composite construct was transplanted into a mouse subcutaneously for six weeks. It resulted in a highly organized matrix, which was similar to native pulp tissue. Another group used DPSCs and collagen with angiogenic growth factors, and introduced this combination into a pulpless root canal for 2 months. A structure of revascularization and tissue regeneration were demonstrated [148]. In addition, Ravindran *et al.* used DPSCs and PDLSCs together in a combined collagen/chitosan scaffold for dental pulp tissue engineering [149]. Chitosan is another natural polymer with antimicrobial property that is favorable in tissue engineering. These hybrid scaffolds seeded with DPSCs and PDLSCs were implanted subcutaneously. Results showed that the constructs developed a pulp-like tissue where cells expressed dentin sialoprotein (DSP) and dentin phosphophoryn (DPP). Furthermore, Erisken *et al.* investigated the viscoelastic properties of dental pulp tissue in an animal model. The purpose of their study was to explore more suitable materials that mimic the microenvironment properties of pulp tissue for regeneration purposes [150]. A comparison was done between natural agarose, alginate, and collagen with native dental pulp tissue. It was found that collagen was the most representative to pulp's viscoelastic behavior and agarose possessed the closest biomechanical property to the pulp tissue. Such findings can be useful when designing complex multi-material based scaffolds. In another study, DPSCs isolated from human subjects were combined with hyaluronan (HA) scaffolds to produce a functional dental pulp-like tissue *in vitro* [151]. Since HA is a natural ECM biomaterial, it aids in the sustainability of homeostasis in different tissues. Also, it is expressed by dental pulp tissue during tooth development. HA scaffolds seeded with DPSCs in combination with differentiation growth factors resulted in dental pulp-like structure *in vitro*. This highlights the endothelial, neural and osteogenic potentials of DPSCs. Another study tested the *in vitro* performance of HA/chitosan combined scaffolds seeded with MSCs for pulp regeneration. It demonstrated good adhesion and proliferation profiles with non-toxic by-products; allowing it to be a safe potential scaffold [152]. Furthermore, Yang *et al.* used another natural polymer, silk fibroin as a scaffold material, and combined it with DPSCs and bFGF [153]. Pulp-like tissue formation occurred when porous silk-based scaffolds, generated via freeze-drying method, combined with bFGF and seeded with DPSCs during *in vitro* and ectopic transplantation tests.

Moreover, Huang *et al.* utilized synthetic copolymer poly-D,L-lactide/glycolide (PLGA) scaffolds for regeneration of dental pulp by seeding them with DPSCs and SCAPs. PLGA has good mechanical properties and is more resistant to contraction than natural polymers such as collagen. Porous PLGA scaffolds were fabricated, via a gas foaming/particulate leaching method, and cell-loaded scaffolds were transferred into the emptied root canal in a mouse model. Observations verified that the empty pulp chamber was filled with vascularized pulp-like structure and dentin-like structure as well [50]. Likewise, synthetic polymer poly-L-lactic acid (PLLA) scaffold was seeded with SHED and implanted within human tooth slices to be inserted into immunodeficient mice. Vascularized pulp-like structures were successfully obtained [154]. Another synthetic polymer of PCL combined with gelatin and composite material was used to fabricate electrospun fibrous scaffolds with nano-hydroxyapatite (nHAp). The hybrid scaffold was seeded with DPSCs, which showed that DPSCs differentiated into odontoblast-like phenotype *in vitro* and *in vivo* [155]. In another study, a composite sandwich-like scaffold consisting of treated dentin matrix (TDM), native dental pulp extracellular matrix (DPEM), and aligned PLGA/gelatin electrospun sheet (APES) were seeded with DFSCs for root regeneration. DPEM/TDM were designed as a differentiation inducer for pulp-dentin part and APES/TDM for the periodontium part [155]. Cell-seeded composite scaffold was transplanted into a miniature swine jaw for 3 months in second premolar extraction sockets. Data revealed that odontoblasts-like layer was developed in the interface of newly formed pre-dentin matrix, as well as dental pulp-like structures with incorporated blood vessels. In addition, TDM surface stimulated the production of cellular cementum and PDL-like structures, which confirmed the suitability of the proposed complex scaffold system for root regeneration.

Bioactive Signals (Bioregulators) for Dental Pulp Tissue Engineering

As a general principle, bioactive signals (bioregulators) such as growth factors, stimulating factors, and cytokines are the third component of tissue engineering strategy along with cells and scaffolds. In dental pulp tissue engineering, several growth factors have been utilized including various subgroups of bone morphogenetic proteins (BMPs), fibroblast growth factors (FGFs), and platelet-derived growth factors (PDGFs) [136].

BMPs play an important role in the regeneration of most of dental tissues in adults by promoting mitogenesis and differentiation. BMPs are dimeric molecules bound with single intermolecular disulfide bond, consisting of four different subgroups, BMP-2 and -4; BMP-3 and BMP-3B (also known as growth/differentiation factor 10 (GDF10); BMPs -5, -6, -7 and -8; GDFs -5, -6 and -7 (also known as cartilage-derived morphogenetic proteins 1, 2 and 3) [141]. *In vitro* and *in vivo* experiments reported that dentin-derived BMP-2 was needed to differentiate SHEDs into odontoblasts [156]. In a monkey model, it was demonstrated that BMP-7 implantation had a role in dentinogenesis. When used in combination with collagen matrix, it becomes a clinically effective pulp-capping agent [157]. In another study, BMP-7 used with collagen/gelatin matrix induced osteodentin formation in the pulp and enabled mineralization. Also, the same study reported the effect of bone sialoprotein (BSP) stimulating coronal pulp mineralization [158]. A similar effect of BSP was observed in a rat model when the carrier-attached BSP was implanted in the pulp of the maxillary first molars [159]. Granulocyte-colony stimulating factor (G-CSF) is known for its migratory effect on DPSCs. When autologous DPSCs combined with G-CSF were transplanted into pulpectomized dog's teeth, pulp tissue regeneration, dentin, and neurovasculature formation occurred [160]. Additionally, some other properties of DPSCs and G-CSF such as antiapoptosis and immunosuppression effects were also evaluated *in vitro* [141].

Another bioactive factor, insulin-like growth factor I (IGF-1), was utilized for pulp healing and dentinogenesis after pulp capping therapy in a rat model. It showed increase in dentinal bridging and tubular dentin formation [161]. The signaling effect of bFGF is well established for inductive wound healing properties in clinical practice. It is also reported to have proliferative effects on DPSCs *in vitro* [162]. In a study, bFGF promoted recellularization and re-vascularization in human teeth, which were endodontically treated and then implanted into the dorsum of rats [163]. In another study, endodontically treated human teeth were implanted in a mouse-model dorsum within collagen scaffolds releasing bFGF and/or vascular endothelial growth factor (VEGF). Findings of recellularization and revascularization of new tissue were observed within the tested specimens [164]. Moreover, combined delivery of bFGF, VEGF or PDGF with a basal set of nerve growth factor (NGF) and BMP-7 exhibited neo-dentin formation. This study successfully utilized various growth factors, implementing the cell homing method instead of transplanting stem cells. Furthermore, FGF-2 is well known for its role in both physiological and pathological conditions. FGF-2 has also been used as a direct infusion, or within a scaffold (hydrogel or collagen sponge), placed into dentinal defects of an amputated pulp in a rat model. Data revealed that non-controlled release of FGF-2 stimulated dentin formation in the residual part of the dental pulp. On the other hand, controlled release of FGF-2 from gelatin hydrogels increased the formation of osteodentin in the pulp. It also increased the formation of dentin bridge-like osteodentin around the regenerated pulp [165]. Thus, developing an effective controlled release system for various bioactive molecules facilitates complex tissue regeneration [136]. However, the dosage, timing, and diverse-delivery control with sustainability of growth factors remain a challenge, which restricts their translational capacity [138]. Gene therapy, however, may provide alternative approaches for such limitations of growth factors [139].

Although dental pulp tissue regeneration seems simpler than regenerating other tissues, there are major regenerative limitations such as its anatomical location, complex histology, and organization [136, 137]. Nevertheless, dental pulp tissue engineering literature presents promising achievements in translational research that could be applied in the near future.

PERIODONTAL TISSUE ENGINEERING AND REGENERATION

The periodontium surrounds and supports the tooth within the alveolar bone. It consists of four different parts, cementum, periodontal ligaments (PDLs), alveolar bone, and gingiva. PDLs are the major component of the periodontium. They consist of vascular bundles of connective tissue with a rich network of collagen fibers. PDLs occupy the area located between the root and the alveolar bone [53 - 56]. Periodontal diseases have a multifactorial etiology involving microbial, genetic, habitual, local, and systemic factors [166]. Periodontitis is a widespread infectious inflammatory disease of alveolar bone. It has a high prevalence and leads to the destruction of the periodontium. As being the most common cause of tooth loss, periodontitis may result in tooth loss if not treated. It is also related to several systemic diseases such as, cancer, rheumatoid arthritis and cardiovascular diseases. Thus, periodontal diseases are one of the most common issues in dentistry. There are some conventional surgical and non-surgical therapies to manage different stages of periodontitis. For example, gingival/ bone grafting, and surgical tissue correction aim to restore functional structure of the damaged periodontium. However, periodontal tissue regeneration is challenging as the periodontium has a complex structure including both soft (PDLs, gingiva) and hard (cementum, alveolar bone) tissue components of various compositions [167 - 170]. As an alternative to current therapies, periodontal tissue engineering, which utilizes stem cells, 3D scaffolds, and growth factors, may offer a chance to improve periodontal regeneration. This will allow the regeneration process to be predictable, qualitative, and less invasive than conventional procedures [168]. Hence, regeneration of the periodontium has been an area of interest and advancement in dental tissue engineering over the last few decades [171]. Although clinical practice has implemented numbers of tissue engineering principles, it is still in its early stages due to complexity of the tissue and dynamics of the oral structures. Most efforts are focusing on *in vivo* periodontal tissue engineering strategies, where stem cells, bioregulators, and scaffolds, or a combination of them, are used in the defect area; as opposed to creating *ex vivo* or *in vitro* artificial tissue constructs [168].

Stem Cells for Periodontal Tissue Engineering

Periodontal cell therapy includes the treatment of periodontitis or other related periodontal diseases by the direct transfer of stem cells into a defect site or via a scaffold to improve the repair process. The transplanted cells may participate in the repair of the damaged periodontium, as structure units. They differentiate into multiple cell types and stimulate tissue recovery by releasing tissue-specific signaling molecules [169]. Several stem cells, isolated from the orofacial region or from other sites of the body, have been utilized for the treatment and regeneration of the periodontium. However, it is clear that the dental stem cells, specifically, periodontal stem cells, are the best fit for periodontium regeneration. Thus, PDLSCs are proposed to be the most promising dental stem cells for the regeneration of periodontal tissue components [53]. *In vivo* studies of a porcine model, autologous [81] or allogeneic [91] PDLSCs were derived from extracted teeth and proliferated *ex vivo*. Then transferred into the periodontal defect zone. It resulted in regeneration of bone, cementum, and PDLs. Similarly, in a rodent model, PDLSCs were transplanted subcutaneously into the dorsal surfaces and periodontal region. PDLSCs aided in the production of a cementum/PDL-like structure [60] and osteogenic tissue [65], which contributed in periodontal tissue repair. Furthermore, in a dog study, PDLSCs were used as cell sheets and transplanted into the periodontal defect. PDLSCs sheets induced periodontal tissue healing with formation of bone, PDL and cementum [87]. Furthermore, Sonoyama *et al.* used PDLSCs in combination with SCAPs to obtain a root periodontal complex. Results suggested that utilizing combined stem cell types improved tissue regeneration *in vivo* [83]. DPSCs were not reported for periodontal regeneration capabilities but they possess high osteogenic potential. This makes DPSCs a valuable source for alveolar bone reconstruction [169]. On the other hand, DFSCs, another type of dental stem cells, were reported to generate PDL-like tissues *in vivo* [113, 172] and can be utilized for periodontal tissue engineering studies.

Adult MSCs from non-dental origin were also tested for the regeneration of periodontal defects. ADSCs, an abundant and accessible stem cell source with low morbidity, were used in periodontal tissue engineering applications as well. In a dog study, it was reported that transplanted ADSCs improved periodontal tissue regeneration [173]. In a rat study, ADSCs were used in combination with platelet-rich plasma (PRP) for periodontal tissue defects repair [174]. After two months, periodontal tissue, including alveolar bone, regeneration was observed. In a rabbit model, Hung *et al.* compared periodontal regenerative capacity of ADSCs with DPSCs. Similar results were found for tooth regeneration *in vivo* and differentiation *in vitro* [2]. In both ADSCs and DPSCs implanted models, they observed tooth-like structures including dentin surrounded by a layer of PDL. They concluded that ADSCs were as useful as DPSCs for periodontal tissue engineering. Another group utilized BMSCs as another stem cell source for periodontal tissue regeneration in a rat model [170]. They encapsulated BMSCs in gelatin micro-carriers and transferred them into an induced periodontal defect area. Regeneration of bone, cementum, and PDL-like structures were achieved within three weeks. Moreover, a significantly higher bone formation and more oriented PDL fibers were obtained compared to the control group that indicated functional repair. Kawaguchi *et al.* examined the periodontal tissue regeneration potential of transplanted autologous BMSCs into periodontal osseous defects in a dog model [175]. Autologous BMSCs were expanded *in vitro* and blended with atelocollagen polymer before transplantation. After one month, they observed cementum, PDL, and alveolar bone regeneration in the defect area. On the other hand, Seo *et al.* reported that BMSCs exhibited less growth than that of DPSCs and PDLSCs *in vitro* [60]. These results were related to the bone marrow origin of BMSCs and that BMSCs might require bone marrow-associated growth factors or other microenvironment components to have continuous proliferation. As an alternative stem cell source, human embryonic stem cells (ESCs) were tested for periodontal tissue engineering due to their unlimited proliferation and differentiation ability [176]. ESCs were cultured in the presence of PDL fibroblastic cells on the extracted tooth root slices. Results revealed that the differentiation of ESCs was stimulated by both root surface and PDL fibroblastic cells. These results seem promising for the use of ESCs in periodontal tissue engineering applications. However, none of these cell transplantation approaches enabled researchers to completely regenerate lost periodontal tissues.

Scaffolds for Periodontal Tissue Engineering

In general, 3D spatial organization is required for most of the regenerative and tissue engineering applications, therefore, the use of scaffolds have been commonly applied. Combining different types of supportive scaffolds with stem cells were evaluated in pre-clinical models for periodontal tissue regeneration [166]. Polymeric biodegradable scaffolds that have porous structures can be fabricated from natural or synthetic materials, such as films, fibers, sheets, gels, or sponges [177]. In animal and clinical studies, β-TCP has been commonly used as a bone graft material with good biocompatibility and osteoconductivity reports. In a rabbit study, β-TCP scaffolds were utilized to expand PDLSCs before transplanting them into periodontal bone defects [178]. It was found that PDLSCs in combination with β-TCP scaffolds contributed in new periodontal bone formation; aiding in periodontal tissue regeneration. In a mouse study, β-TCP mixed with hydroxyapatite HAp particles were seeded with PDLSCs before transplantation. It resulted in a cementum-like tissue formation associated with PDL tissue [82]. HAp, another scaffolding material, of a natural polymer, was utilized as a carrier and combined with PDLSCs to repair periodontal defects of bilateral mandibular first molars, in a dog model [87]. In this study, PDLSCs were applied as cell sheets that resulted in periodontal tissue healing with new bone, PDL, and cementum formation. By using similar cell sheets approach, another group utilized utilized calcium phosphate-coated PCL scaffolds as a supportive as a supportive material for the cells isolated from human PDL, alveolar bone, and gingiva [167]. Cell-laden scaffolds were transferred into rat periodontal defects. Findings indicated that the most effective cell sheet scaffold complex was PDLSCs-based regeneration. And the promotion of periodontal regeneration was significantly higher than the other cell-seeded or unseeded-scaffolds. Moreover, *in vitro* performance of honeycomb shaped polymeric PCL films was investigated by seeding them with PDL cells isolated from extracted molars [178]. The porous honeycomb shaped films promoted multi-layered cell sheet formation and PDL cell differentiation more than with flat PCL films. Another group of researchers fabricated more precise fibrous PCL scaffolds using solid freeform method. This form of scaffold can anatomically fit complex defects with enhanced biomimetic architecture [179]. Human PDL cells were seeded on scaffolds prior to transplantation into the defect areas of a rat model. Periostin, a marker for PDL tissue maturation, demonstrated a pattern distribution in the healing area after three and six weeks, which implied functional periodontal regeneration. Moreover, periostin activity in the oriented-fibrous scaffolds was significantly higher than that of random porous scaffolds, which makes it more suitable for long-term maintenance of engineered periodontal tissue.

Poly(lactic-co-glycolic acid (PLGA) is another synthetic copolymer that is used as a scaffolding material. PLGA scaffolds were combined with autologous PDLSCs and genetically-modified BMSCs co-culture, which were utilized to treat periodontal bone defects in a dog model [180]. Formation of new alveolar bone, cementum, and connective tissue in the cell laden PLGA scaffolds implanted sites were significantly higher than that of the control group. Additionally, BMSCs were highly expressive of osteoprotegerin. Thus, these scaffold-cells combination and gene therapy might be useful for periodontal tissue engineering applications. Similarly, *in vitro* PDL regeneration of non-woven 3D polyglycolic acid (PGA) constructs seeded with PDL cells was investigated. Cells adhered well to the scaffolds and displayed physiological ECM secretion [181]. The use of nHAp could be another potential strategy for periodontal tissue regeneration. The size and surface effects at the nanosized nHAp particles have unique properties over their bulk-phase counterparts. One group investigated the effects of nHAp particles on human PDLSCs *in vitro*. It was found that nHAp significantly increased BMP-2 expression. This could imply that the nano-sized nHAp might regulate PDLSCs differentiation [178]. Since PDL has a key role in tooth anchorage, tissue engineered PDL on titanium dental implants have been investigated. In a dog study, PDL cells were cultured in a bioreactor on titanium pins [182]. After the formation of multiple cellular layers, titanium pins were transplanted into alveolar bone PDL-like tissue formation on the surface of dental implants was identified, highlighting the principle of ligament-anchored implants.

Bioactive Signals (Bioregulators) for Periodontal Tissue Engineering

Growth factor signaling stimulates and regulates different biological processes of cells including proliferation and differentiation of stem cells. As previously described, growth factors are a valuable and efficient component of tissue engineering. Numerous bioactive signals (bioregulators), such as PDGF, TGF, FGF family and BMP family, have been proposed to promote the mechanism of periodontal regeneration [166]. In most periodontal regeneration studies, the growth factors were applied directly or through a delivery system into the defect area. On the other hand, future tissue engineering approaches for periodontal regeneration consist of controlled release of growth factors from scaffolds, with stem cells, and a proper microenvironment for cellular development [177].

The action mechanism of FGF-2 (bFGF) during periodontal regeneration was investigated in a dog model. FGF-2, dissolved in hydroxypropyl cellulose solution, was delivered to surgically made periodontal defects. The results revealed that the proliferation of bone marrow and PDL derived fibroblastic cells were initially enhanced by FGF-2. Then, there was improved blood vessel development and growth, and lastly, bone formation with osteoblastic differentiation. Therefore, FGF-2 accelerated new tissue formation at the early phase of regeneration that resulted in PDL, cementum, and new bone formation [183]. In an *in vitro* study, it was reported that FGF-2 was highly effective on enhancing the adhesive and proliferative capacity of PDLSCs. Moreover, FGF-2 and VEGF possessed a synergistic effect on PDLSCs [184]. In a dog study, it was reported that the topical application of bFGF in alveolar bone defects, significantly induced PDL regeneration without ankylosis, root resorption, and epithelial down growth [185]. Similar results were reported in a randomized multicenter clinical trial [186], and in a dose dependent manner [187]. Transforming growth factor-β1 (TGF-β1) influences cellular activates on various levels, such as the control of cellular growth and apoptosis. However, there are a few studies reporting the effects of TGF-β1 on PDL cells. Fujii *et al.* investigated the expression of TGF-β1 in PDL in rats. They also studied the exogenous TGF-β1 effects on the proliferation and gene expression in human PDL cells and PDLSCs [188]. It was found that application of TGF-β1 stimulated fibroblastic differentiation of PDLSCs and maintenance of the PDL complex. In addition, they demonstrated that both PDL cells and PDLSCs expressed TGF-β1. It was also reported that TGF-β enhanced surface proteoglycan genes of PDL cells such as syndecan-2 and β-glycan *in vitro* [189].

Moreover, gene therapy presents a promising role in tissue engineering and regenerative medicine. Utilizing cells transfected with growth factor genes are a possible modality for periodontal tissue engineering. Li *et al.* used BMP-7 gene transfected BMSCs for the treatment of periodontal defects in a dog model [190]. BMSCs were seeded in collagen matrices and transferred into defect areas. Results demonstrated that in addition to the alveolar bone regeneration, the amount of new cementum formation was significantly enhanced by BMP-7 gene transfected BMSCs. Another study showed similar results for BMP-2 gene-transfected BMSCs. Among the various growth factors used in periodontal regeneration *in vitro* and *in vivo*, PDGF (GEM21S1) and BMP-2 (Infuse1) are commercially available and in clinical use in the United States [177].

Periodontal regeneration is thought to be challenging in dentistry because of the complexity and diversity of the tissue, as well as the oral microbial involvement. Therefore, the development of tissue engineering based therapies will open new doors for periodontal regeneration. The stem cells of dental/ oral origin and other sources of MSCs have displayed promising results in the regeneration of periodontal defects. In addition, several scaffold materials and growth factors have been utilized for such purposes. Combining the potential advantages of stem cells, scaffold biomaterials, and growth factors would be more favorable to achieve the ultimate goal of periodontal tissue regeneration.

TISSUE ENGINEERING FOR PERIODONTAL INTROBONY DEFECTS

As a part of the peridontium, alveolar bone takes role in structural support and maintenance of teeth. Orofacial bones are subjected to critical stress and strain produced by different muscles during mastication. Therefore, different responses to biological and mechanical stimuli are generated in different parts of the masticatory system. Thus, regeneration of these tissues is challenging and requires synergy of both the cellular and molecular activities [3]. Periodontitis can affect all periodontal tissue components, as explained in the section "Periodontal Tissue Engineering and Regeneration", and can result in pathological changes in the alveolar bone morphology. Such bone loss, due to severe infection, is mostly bacterial-induced that consequently leads to tooth loss [191]. Other causes of bone loss, such as trauma, could lead to soft and hard tissue defect as well [192]. An intrabony defect or an angular defect is described as a periodontal defect within bone, circumvented by one, two, or three bony walls, or a combination of them [193]. Several treatments, such as bone grafts and guided tissue regeneration with physical barriers have been approved for clinical use. However, each modality has its own limitations such as insufficient autologous graft tissue, limited pool of site cells, and donor site morbidity. Tissue engineering may be an alternative and effective strategy to treat these clinical concerns [194].

Stem Cells for Intrabony Defects

Healthy PDL tissues host stem cells throughout adulthood that enable the periodontium and the surrounding bone tissues to maintain their regenerative capacity. On the other hand, due to the negative effects of advanced periodontitis or other related problems on stem cell niches, tissue regeneration is impaired because of the functional limitations of stem cells. Since new tissue formation resulted from cellular activity, delivery of inherently capable cells, especially stem cells, into the defect area has been proposed to regenerate tissue. Therefore, *in vitro* expanded stem cells are suggested to be an alternative source to achieve periodontal bone regeneration [3, 55].

As stated in the previous sections of this chapter, DPSCs have a great potential for various tissue-engineering applications. They displayed promising results for the treatment of angular bony defects. Jahanbin *et al.* used human DPSCs for the regeneration of maxillary alveolar bony defects, in a rat model [195]. They assessed the treatment performance of DPSCs with BMSCs. Results revealed that both groups showed a promising outcomes for the regeneration of alveolar bony defects. In addition, bone formation of BMSCs was higher than that of DPSCs after eight weeks. Similarly, Rai *et al.* utilized DPSCs for intrabony defects where the cells were delivered within a scaffold, which resulted in a successful bone formation [196]. In another rat study, BMSCs were employed for both alveolar bone and soft tissue regeneration [170]. Green fluorescent protein (GFP) -labeled rat BMSCs were cultured within gelatin carriers and transplanted into the defect areas. Three weeks after transplantation, soft and hard tissue regeneration were evident in both the experimental and control groups. However, new bone formation was significantly higher in the experimental group. Furthermore, it was observed that GFP-positive BMSCs were integrated into the new bone parts, which indicated the direct contribution of transplanted BMSCs to tissue regeneration. In a canine model study, it was found that BMSCs and alveolar periosteal cells sheets were able to regenerate periodontal tissue in a one wall intrabony defect [197]. Nonetheless, the newly formed cementum and oriented PDL fibers were less than that of the PDLSCs sheet group. Another group of researchers used autologous PDLSCs to treat periodontitis and intrabony defects in a miniature porcine model [81]. Alveolar bone defects were instrumentally created around the maxillary and mandibular first molars. PDLSCs were labeled with GFP for tracking and identification purposes. After three months, PDLSCs were identified in the newly formed bone, as osteoblasts. Results showed that PDLSCs displayed a good potential to generate bone, cementum and PDL.

Scaffolds for Intrabony Defects

In addition to standard bone grafting methods, there are several natural and synthetic materials that are used as scaffold materials in orofacial bone regeneration. Scaffolds that are architecturally or biologically suitable for bone regeneration are generally based on one or more of the naturally occurring bone components, such as collagen, and inorganic bone components, such as hydroxyapatite (HAp) [191].

HAp and β-TCP have been reported among the most widely used biocompatible bone grafting materials for the repair of periodontal bony defects and alveolar bone grafting procedures [198]. More than half of the human hard tissue mineral component is HAp and it consists of calcium phosphate aggregates. Thus, HAp is an ideal scaffolding material to mimic bone mineral composition. TCP has a similar chemical composition to HAp, and can be degraded during new bone formation better than Hap [177]. In a clinical study, researchers compared the performance of HAp and β-TCP for periodontal defect therapy. After six months, findings revealed that the HAp group displayed better results than the β-TCP group, but with no statistical significance. Thus both materials provided beneficial outcomes for the treatment of periodontal bony defects. In another study, bilayered constructs were developed for the regeneration of both soft and hard tissue components of the periodontium [199]. The scaffold for the bone part was fabricated from PCL and β-TCP mixed polymer via fused deposition method. The

PDL part, was electrospun PCL membrane where osteoblasts and PDL cell sheets were incorporated within the scaffolds respectively. *In vitro* results revealed mineralized matrix formation at three weeks. *In vivo* results in a rat, subcutaneous, model supported the *in vitro* data and the combined scaffold system stimulated the regeneration of alveolar bone and PDL tissues. In another study, PLGA sponge scaffolds were evaluated for their periodontal bone regeneration capacity [200]. Scaffolds were seeded with cementoblasts, PDL fibroblasts, or DF cells. Overall *in vitro* and *in vivo* results suggested that all cell-seeded PLGA scaffolds could promote periodontal reconstruction. Natural polymers are another alternative scaffold material. Main benefits of natural polymers, such as collagen, are their biocompatibility, biodegradability, and ability to bind to growth factors needed for osteoinduction. Collagen is the most commercially available natural polymer for periodontal bone regeneration applications [191]. A clinical study examined collagen sponge scaffolds with autologous DPSCs for the treatment of periodontal intrabony defects, which resulted from chronic periodontitis [196]. In this study, collagen was chosen as the scaffold material, as it is the major protein found in many tissues such as undifferentiated mesenchymal tissue and dentin. It also can support stem cell differentiation of dentinogenesis. Third molars were utilized as the source of DPSCs, where they were isolated and transferred into the intrabony defects within the collagen scaffolds. Results in one year demonstrated that there was no grafting detachment at any site and all the defects were covered with bone like tissue.

For the first time as a direct application of PDLSCs in a clinical study, Vandana *et al.* employed autologous PDLSCs in the treatment of intrabony defects [201]. This method enabled researchers to address the problems of *ex vivo* stem cell culture procedures such as loss of stemness in the cell passaging period, high costs, genetic manipulation, and possible tetratomic effects. In this clinical trial, PDLSCs were isolated from impacted third molars and loaded into a gelatin based sponge, Abgel®, which was used as a scaffold. Cementum scrapings (which include several bioactive molecules such as VEGF) were used as signaling agents. One year after delivery decreased periodontal pocket depth and sufficient intrabony defect fill were found. This clinical study demonstrated the therapeutic potential of this new technique. Gelfoam® is another gelatin based sponge which was utilized in dental bony defect therapy [202]. In another clinical study, a combination of a pre-formed bone minerals and collagen (Bio-Oss® Collagen) was used as a commercially available bone grafting material. It was demonstrated that the material facilitated the regeneration of the periodontal intrabony defects [203]. In a dog study, a β-TCP based commercially available material, Cerasorb®, was seeded and incubated with BMSCs for intrabony defect treatment [204]. Results implied that this material stimulated stem cell proliferation and osteogenic marker production. After six months, it was observed that Cerasorb® was completely replaced by new bone in the intrabony defect area. Coral derived porous 3D HAp scaffold, Biocoral®, was another commercial material, which was evaluated as a bone replacement graft in human intrabony defects [205]. Five-year results revealed that the material enabled significant clinical improvement and it was beneficial for periodontal bone defect therapy.

Moreover, there are several collagen-containing commercial membranes such as Ossix™, Bio-Gide®, Osteovit®, Neomem®, Biomend™, and Biomend Extend™ [3, 202]. These membranes can be used directly as a framework or by seeding with stem cells. Such membranes can be transferred into the defect area as viable cell containing matrices. There are also other synthetic polymers used as scaffolding materials in commercial periodontal use, such as polytetrafluoroethylene (Gore-Tex®), PLA (Vivosorb® and Epi-Guide®), and PGA (Gore Resolut Adapt®) [3].

Bioactive Signals (Bioregulators) for Intrabony Defects

Stimulation of osteogenesis, cementogenesis, and new tissue formation via the delivery of various signaling molecules are necessary for periodontal tissue engineering and intrabony defect therapy. Several recombinant human growth factors such as BMP and PDGF have been positively reported in pre-clinical and clinical trials. Biological agents, such as enamel matrix derivatives (EMD) and biologically active products derived from patients, such as PRP, have also been reported. These bio-regulators are currently used as clinical tools, or are under investigation in controlled clinical studies for tissue regeneration purposes [169].

Platelet-rich fibrin (PRF) is an individually extracted aggregate, used in periodontal regeneration therapy and other treatments as well. It is a fibrin matrix enriched with various bioactive factors such as PDGF, VEGF, TGF, IGF, EGF, and bFGF. It also contains viable platelets, which can be obtained from an anticoagulant-free blood harvest. Naturally forming PRF possesses a dense 3D architecture that makes it a good scaffold candidate. PRF stimulates the proliferation of osteoblasts, gingival fibroblasts, and PDL cells. It suppresses the activity of epithelial cells, therefore, it can be geared toward the proliferation of a select tissue of interest such as bone [192, 206 - 208]. On the other hand, PRF lacks rigidity to be able to support bony defects, particularly, the load-bearing area. Thus, it is often blended with mineralized bone grafts. A clinical study utilized PRF for alveolar bone regeneration in intrabony defects [207]. PRF was mixed with β-TCP and then transferred into defects in the mandibular region. After six months, results showed reduced periodontal probing depths, increased clinical attachment levels, gingival thickness and periapical bone density. In another study, HAp particles were utilized as a bone substitute for dental bone defects in combination with PRF membrane [192]. Such combination was considered because PRF can provide a concentrated suspension of growth factors found in platelets. Results presented rapid healing, good attachment gain in the defect area, and a decreased periodontal pocket depth. Similar clinical and radiographic results were obtained in a case study using PRF mixed with a commercial alloplast bone graft material, Ossifi™ [206]. Improved patient comfort and wound healing were also reported. In another report, alveolar bone regeneration was studied around osseointegrated dental implants [208]. Implants were coated with HAp and transplanted into the defect areas following the delivery of allograft stem cells (consisting of PRP with either DPSCs or BMSCs). The PRP represented a niche for growth factors such as PDGF, TGF-β1, and IGF. Results showed that both stem cell types have the ability to induce bone formation when applied with PRP, which improves bone-implant interaction.

As a third generation periodontal regeneration therapy tool, EMD (commercially available as Emdogain®,), contains more than 90% amelogenin and other proteins. It was reported for periodontal and regenerative use [3]. EMD has a stimulatory effect on the proliferation and differentiation of human PDL cells, and amelogenin can selectively trigger human PDL fibroblasts. Moreover, EMD acts as a proangiogenic factor and promotes blood vessel formation during periodontal regeneration. Five-year follow-up results demonstrated an improved clinical attachment level and bone fill, which supports the use of EMD for the treatment of intrabony defects [209].

It is known that the BMP/TGF-β signaling pathway stimulates osteoblastic differentiation and *in vivo* bone formation. Current literature reveals that BMP-2, BMP-4, BMP-7, and BMP-12 are potent mediators for alveolar bone formation [194]. Thus, the BMP family and particularly, BMP-2, has been extensively studied for periodontal regeneration. One group studied the effect of BMP--mediated DPSCs on the regeneration of alveolar bone defects in a rabbit model [210]. NHAp/collagen/PLA blend scaffold was used for an effective delivery of autologous stem cells and growth factors into the defect. The generated bone complex demonstrated an earlier mineralization and more bone formation compared to control groups. Results implied that BMP-2 promoted osteogenic capacity of DPSCs and the scaffold served as an optimal template for seeding, proliferation, and differentiation of the DPSCs. The half-life of the growth factors *in vivo* is short. Thus, gene therapy may present sustainable release of growth factors within the periodontal defects and provide higher potential for regeneration [55]. In a beagle dog study, genetically modified BMP-2 expressing MSCs were evaluated for mandibular bone regeneration [211]. Alveolar bone defects were surgically created at premolars sites, and BMP-2 modified or native MSCs (control) were implanted. Two months after application, the experimental group displayed significant increase in bone regeneration than the control group, which indicated its use in clinical bone repair. Other members of the BMP family, such as BMP-7, BMP-12 and BMP-14 were also assessed in animal models for their efficiency in periodontal regeneration [194]. BMP-2 (Infuse®), BMP-7 (OP-1®), and PDGF (GEM21S®) have already become commercially available growth factors for clinical use in the United States. In addition, FGF-2 has been investigated in a clinical Phase III trial in Japan with some of the results reported [177, 202]. Moreover, the effect of stromal cell-derived factor-1 (SDF-1) on periodontal bone regeneration was investigated in a rat model [212]. SDF-1 is a chemokine that is crucial for stem cell recruitment in tissue repair. SDF-1 was delivered via a collagen scaffold that was inserted into the mandibular bone defect. Their findings suggested that SDF-1 promoted the migration of host MSCs into the defect area, reduced the inflammatory response, induced early osteogenesis, and vascularization, as well as it increased the quality and quantity of the regenerated bone tissue. In experimental setting, it is known that MSCs release their cytokines such as VEGF, TGF, and IGF to the ECM or the culture medium [212].

Despite the remarkable advances of tissue engineering in dentistry, the complete regeneration of periodontal bone tissues is still challenging. Tissue engineering strategies using scaffolds, growth factors, stem cells, and gene therapy possess the potentials to overcome many of the current limitations. In order to cope with the therapeutic limitations and generate stem cell based strategies, it is important to optimize cell scaffold incorporation by understanding the molecular, cellular, and functional activities during periodontal therapy and regeneration [177, 191].

CONCLUSION

Tissue engineering approach combines the potential advantages of stem cells, scaffold biomaterials, and growth factors to achieve the ultimate goal of tissue regeneration. Dental tissue engineering has drastically improved, yet it is still in its early stages with several challenges to address. It is necessary to consider an effective strategy capable of producing scaffolds with the correct physical, biological, and mechanical features. Most importantly, it is pivotal to integrate the nature and dynamics of the masticatory system and oral environment with the discussed tissue engineering principles, to implement the transitional goals and strategies.

ABBREVATIONS

ADSCs   adipose tissue derived stem cells

APES   aligned PLGA/gelatin electrospun sheet

bFGF   basic fibroblast growth factor

BMP	bone morphogenetic protein
BMSCs	bone marrow stem cells
BSP	bone sialoprotein
β-TCP	beta-tricalcium phosphate
DF	dental follicle
DFSCs	dental follicle stem cells
DMP1	dentin matrix protein 1
DPEM	dental pulp extracellular matrix
DPP	dentin phosphophoryn
DPSCs	dental pulp stem cells
DSP	dentin sialoprotein
ECM	extracellular matrix
EMD	enamel matrix derivative
ESCs	embryonic stem cells
G-CSF	granulocyte-colony stimulating factor
GFP	green fluorescent protein
HA	hyaluronan
HAp	hydroxyapatite
IGF	insulin-like growth factor
MSCs	mesenchymal stem cells
NGF	nerve growth factor
nHAp	nano-hydroxyapatite
PAFSCs	periapical follicle stem cells
PCL	poly-epsilon-caprolactone
PDGF	platelet-derived growth factor
PDLs	periodontal ligaments
PDLSCs	periodontal ligament stem cells
PGA	polyglycolic acid
PLGA	poly-D,L-lactide-co-glycolide
PLLA	poly-L-lactic acid
PRF	platelet-rich fibrin
PRP	platelet-rich plasma
SCAP	root apical papilla stem cells
SDF-1	stromal cell-derived factor-1
SHED	stem cells from human exfoliated deciduous teeth

TDM	treated dentin matrix
TGF- β1	transforming growth factor-β1
VEGF	vascular endothelial growth factor

CONFLICT OF INTEREST

The authors declare no conflict of interest, financial or otherwise.

ACKNOWLEDGEMENTS

Declared none.

REFERENCES

[1]   Singh K, Mishra N, Kumar L, Agarwal KK, Agarwal B. Role of stem cells in tooth bioengineering. J Oral Biol Craniofac Res 2012; 2(1): 41-5.
        [http://dx.doi.org/10.1016/S2212-4268(12)60010-4] [PMID: 25756031]

[2]   Hung C-N, Mar K, Chang H-C, *et al.* A comparison between adipose tissue and dental pulp as sources of MSCs for tooth regeneration. Biomaterials 2011; 32(29): 6995-7005.
        [http://dx.doi.org/10.1016/j.biomaterials.2011.05.086] [PMID: 21696818]

[3]   Abou Neel EA, Chrzanowski W, Salih VM, Kim H-W, Knowles JC. Tissue engineering in dentistry. J Dent 2014; 42(8): 915-28.
        [http://dx.doi.org/10.1016/j.jdent.2014.05.008] [PMID: 24880036]

[4]   Volponi AA, Pang Y, Sharpe PT. Stem cell-based biological tooth repair and regeneration. Trends Cell Biol 2010; 20(12): 715-22.
        [http://dx.doi.org/10.1016/j.tcb.2010.09.012] [PMID: 21035344]

[5]   Park Y-J, Cha S, Park Y-S. Regenerative applications using tooth derived stem cells in other than tooth regeneration: a literature review. Stem Cells Int 2016; 2016: 1-12.
        [http://dx.doi.org/10.1155/2016/9305986]

[6]   Mao JJ. Stem cells and the future of dental care. N Y State Dent J 2008; 74(2): 20-4.
        [PMID: 18450184]

[7]   Hosoya A, Nakamura H. Ability of stem and progenitor cells in the dental pulp to form hard tissue. Jpn Dent Sci Rev 2015; 51(3): 75-83.
        [http://dx.doi.org/10.1016/j.jdsr.2015.03.002]

[8]   La Noce M, Paino F, Spina A, *et al.* Dental pulp stem cells: state of the art and suggestions for a true translation of research into therapy. J Dent 2014; 42(7): 761-8.
        [http://dx.doi.org/10.1016/j.jdent.2014.02.018] [PMID: 24589847]

[9]   Pashley DH, Walton RE, Slavkin HC. Histology and physiology of the dental pulp. Ingle JI, Bakland LK Endodontics. Hamilton, Ont.: B C Decker 2002; pp. 25-61.

[10]   Gronthos S, Mankani M, Brahim J, Robey PG, Shi S. Postnatal human dental pulp stem cells (DPSCs) *in vitro* and *in vivo*. Proc Natl Acad Sci USA 2000; 97(25): 13625-30.
        [http://dx.doi.org/10.1073/pnas.240309797] [PMID: 11087820]

[11]   Gronthos S, Brahim J, Li W, *et al.* Stem cell properties of human dental pulp stem cells. J Dent Res 2002; 81(8): 531-5.
        [http://dx.doi.org/10.1177/154405910208100806] [PMID: 12147742]

[12]   Ferro F, Spelat R, Baheney CS. Dental Pulp Stem Cell (DPSC) Isolation, Characterization, and Differentiation. In: Kioussi C, Ed. Stem Cells and Tissue Repair. New York, NY: Springer New York

2014; pp. 91-115.
[http://dx.doi.org/10.1007/978-1-4939-1435-7_8]

[13]   Patil R, Kumar BM, Lee W-J, *et al.* Multilineage potential and proteomic profiling of human dental stem cells derived from a single donor. Exp Cell Res 2014; 320(1): 92-107.
[http://dx.doi.org/10.1016/j.yexcr.2013.10.005] [PMID: 24162002]

[14]   Rajendran R, Gopal S, Masood H, Vivek P, Deb K. Regenerative potential of dental pulp mesenchymal stem cells harvested from high caries patient's teeth. J Stem Cells 2013; 8(1): 25-41.
[PMID: 24459811]

[15]   Alge DL, Zhou D, Adams LL, *et al.* Donor-matched comparison of dental pulp stem cells and bone marrow-derived mesenchymal stem cells in a rat model. J Tissue Eng Regen Med 2010; 4(1): 73-81.
[PMID: 19842108]

[16]   Arthur A, Shi S, Gronthos S. Dental Pulp Stem Cells. Stem cell biology and tissue engineering in dental sciences. Elsevier 2015; pp. 279-89.
[http://dx.doi.org/10.1016/B978-0-12-397157-9.00023-0]

[17]   Janebodin K, Horst OV, Ieronimakis N, Balasundaram G, Reesukumal K, Pratumvinit B, *et al.* Isolation and characterization of neural crest-derived stem cells from dental pulp of neonatal mice. In: Shi S, Ed. PLoS ONE 2011; 6(11): e27526.
[http://dx.doi.org/10.1371/journal.pone.0027526]

[18]   Graziano A, d'Aquino R, Laino G, Papaccio G. Dental pulp stem cells: a promising tool for bone regeneration. Stem Cell Rev 2008; 4(1): 21-6.
[http://dx.doi.org/10.1007/s12015-008-9015-3] [PMID: 18300003]

[19]   Pierdomenico L, Bonsi L, Calvitti M, *et al.* Multipotent mesenchymal stem cells with immunosuppressive activity can be easily isolated from dental pulp. Transplantation 2005; 80(6): 836-42.
[http://dx.doi.org/10.1097/01.tp.0000173794.72151.88] [PMID: 16210973]

[20]   Yamaza T, Kentaro A, Chen C, *et al.* Immunomodulatory properties of stem cells from human exfoliated deciduous teeth. Stem Cell Res Ther 2010; 1(1): 5.
[http://dx.doi.org/10.1186/scrt5] [PMID: 20504286]

[21]   Pisciotta A, Carnevale G, Meloni S, *et al.* Human dental pulp stem cells (hDPSCs): isolation, enrichment and comparative differentiation of two sub-populations. BMC Dev Biol 2015; 15(1): 14.
[http://dx.doi.org/10.1186/s12861-015-0065-x] [PMID: 25879198]

[22]   Lizier NF, Kerkis A, Gomes CM, *et al.* Scaling-Up of Dental Pulp Stem Cells Isolated from Multiple Niches. 2012.
[http://dx.doi.org/10.1371/journal.pone.0039885]

[23]   Rai S, Kaur M, Kaur S, Arora SP. Redefining the potential applications of dental stem cells: An asset for future. Indian J Hum Genet 2012; 18(3): 276-84.
[http://dx.doi.org/10.4103/0971-6866.107976] [PMID: 23716933]

[24]   Sonoyama W, Liu Y, Yamaza T, *et al.* Characterization of the apical papilla and its residing stem cells from human immature permanent teeth: a pilot study. J Endod 2008; 34(2): 166-71.
[http://dx.doi.org/10.1016/j.joen.2007.11.021] [PMID: 18215674]

[25]   Takeda T, Tezuka Y, Horiuchi M, *et al.* Characterization of dental pulp stem cells of human tooth germs. J Dent Res 2008; 87(7): 676-81.
[http://dx.doi.org/10.1177/154405910808700716] [PMID: 18573990]

[26]   Atari M, Gil-Recio C, Fabregat M, *et al.* Dental pulp of the third molar: a new source of pluripotent-like stem cells. J Cell Sci 2012; 125(Pt 14): 3343-56.
[http://dx.doi.org/10.1242/jcs.096537] [PMID: 22467856]

[27]   Hosoya A, Nakamura H, Ninomiya T, *et al.* Hard tissue formation in subcutaneously transplanted rat dental pulp. J Dent Res 2007; 86(5): 469-74.

[http://dx.doi.org/10.1177/154405910708600515] [PMID: 17452570]

[28] Kuo TF, Lee S-Y, Wu H-D, Poma M, Wu Y-W, Yang J-C. An *in vivo* swine study for xeno-grafts of calcium sulfate-based bone grafts with human dental pulp stem cells (hDPSCs). Mater Sci Eng C 2015; 50: 19-23.
[http://dx.doi.org/10.1016/j.msec.2015.01.092] [PMID: 25746240]

[29] Petridis X, Diamanti E, Trigas GCh, Kalyvas D, Kitraki E. Bone regeneration in critical-size calvarial defects using human dental pulp cells in an extracellular matrix-based scaffold. J Craniomaxillofac Surg 2015; 43(4): 483-90.
[http://dx.doi.org/10.1016/j.jcms.2015.02.003] [PMID: 25753474]

[30] Yasui T, Mabuchi Y, Toriumi H, *et al.* Purified Human Dental Pulp Stem Cells Promote Osteogenic Regeneration. J Dent Res 2016; 95(2): 206-14.
[http://dx.doi.org/10.1177/0022034515610748] [PMID: 26494655]

[31] Lee J-H, Lee D-S, Choung H-W, *et al.* Odontogenic differentiation of human dental pulp stem cells induced by preameloblast-derived factors. Biomaterials 2011; 32(36): 9696-706.
[http://dx.doi.org/10.1016/j.biomaterials.2011.09.007] [PMID: 21925730]

[32] Dimitrova-Nakov S, Baudry A, Harichane Y, Kellermann O, Goldberg M. Pulp stem cells: implication in reparative dentin formation. J Endod 2014; 40(4) (Suppl.): S13-8.
[http://dx.doi.org/10.1016/j.joen.2014.01.011] [PMID: 24698687]

[33] Arthur A, Rychkov G, Shi S, Koblar SA, Gronthos S. Adult human dental pulp stem cells differentiate toward functionally active neurons under appropriate environmental cues. Stem Cells 2008; 26(7): 1787-95.
[http://dx.doi.org/10.1634/stemcells.2007-0979] [PMID: 18499892]

[34] Ellis KM, O'Carroll DC, Lewis MD, Rychkov GY, Koblar SA. Neurogenic potential of dental pulp stem cells isolated from murine incisors. Stem Cell Res Ther 2014; 5(1): 30.
[http://dx.doi.org/10.1186/scrt419] [PMID: 24572146]

[35] Isobe Y, Koyama N, Nakao K, *et al.* Comparison of human mesenchymal stem cells derived from bone marrow, synovial fluid, adult dental pulp, and exfoliated deciduous tooth pulp. Int J Oral Maxillofac Surg 2016; 45(1): 124-31.
[http://dx.doi.org/10.1016/j.ijom.2015.06.022] [PMID: 26235629]

[36] Rizk A, Rabie AB. Human dental pulp stem cells expressing transforming growth factor $\beta3$ transgene for cartilage-like tissue engineering. Cytotherapy 2013; 15(6): 712-25.
[http://dx.doi.org/10.1016/j.jcyt.2013.01.012] [PMID: 23474328]

[37] Vasanthan P, Gnanasegaran N, Govindasamy V, *et al.* Comparison of fetal bovine serum and human platelet lysate in cultivation and differentiation of dental pulp stem cells into hepatic lineage cells. Biochem Eng J 2014; 88: 142-53.
[http://dx.doi.org/10.1016/j.bej.2014.04.007]

[38] Garzón I, Martin-Piedra MA, Alaminos M. Human dental pulp stem cells. A promising epithelial-like cell source. Med Hypotheses 2015; 84(5): 516-7.
[http://dx.doi.org/10.1016/j.mehy.2015.02.020] [PMID: 25764965]

[39] Bray AF, Cevallos RR, Gazarian K, Lamas M. Human dental pulp stem cells respond to cues from the rat retina and differentiate to express the retinal neuronal marker rhodopsin. Neuroscience 2014; 280: 142-55.
[http://dx.doi.org/10.1016/j.neuroscience.2014.09.023] [PMID: 25242642]

[40] Yam GH, Peh GS, Singhal S, Goh B-T, Mehta JS. Dental stem cells: a future asset of ocular cell therapy. Expert Rev Mol Med 2015; 17: e20.
[http://dx.doi.org/10.1017/erm.2015.16] [PMID: 26553416]

[41] Wakayama H, Hashimoto N, Matsushita Y, *et al.* Factors secreted from dental pulp stem cells show multifaceted benefits for treating acute lung injury in mice. Cytotherapy 2015; 17(8): 1119-29.

[http://dx.doi.org/10.1016/j.jcyt.2015.04.009] [PMID: 26031744]

[42]   Hattori Y, Kim H, Tsuboi N, *et al.* Therapeutic Potential of Stem Cells from Human Exfoliated Deciduous Teeth in Models of Acute Kidney Injury. 2015.

[43]   Kanafi MM, Rajeshwari YB, Gupta S, *et al.* Transplantation of islet-like cell clusters derived from human dental pulp stem cells restores normoglycemia in diabetic mice. Cytotherapy 2013; 15(10): 1228-36.
       [http://dx.doi.org/10.1016/j.jcyt.2013.05.008] [PMID: 23845187]

[44]   Yamaguchi S, Shibata R, Yamamoto N, *et al.* Dental pulp-derived stem cell conditioned medium reduces cardiac injury following ischemia-reperfusion. Sci Rep 2015; 5: 16295.
       [http://dx.doi.org/10.1038/srep16295] [PMID: 26542315]

[45]   Hirose Y, Yamamoto T, Nakashima M, *et al.* Injection of Dental Pulp Stem Cells Promotes Healing of Damaged Bladder Tissue in a Rat Model of Chemically Induced Cystitis. Cell Transplant 2016; 25(3): 425-36.
       [http://dx.doi.org/10.3727/096368915X689523] [PMID: 26395427]

[46]   Griesi-Oliveira K, Sunaga DY, Alvizi L, Vadasz E, Passos-Bueno MR. Stem cells as a good tool to investigate dysregulated biological systems in autism spectrum disorders. Autism Res 2013; 6(5): 354-61.
       [http://dx.doi.org/10.1002/aur.1296] [PMID: 23801657]

[47]   Yu J, Deng Z, Shi J, *et al.* Differentiation of dental pulp stem cells into regular-shaped dentin-pulp complex induced by tooth germ cell conditioned medium. Tissue Eng 2006; 12(11): 3097-105.
       [http://dx.doi.org/10.1089/ten.2006.12.3097] [PMID: 17518625]

[48]   Yu J, Wang Y, Deng Z, *et al.* Odontogenic capability: bone marrow stromal stem cells *versus* dental pulp stem cells. Biol Cell 2007; 99(8): 465-74.
       [http://dx.doi.org/10.1042/BC20070013] [PMID: 17371295]

[49]   Batouli S, Miura M, Brahim J, *et al.* Comparison of stem-cell-mediated osteogenesis and dentinogenesis. J Dent Res 2003; 82(12): 976-81.
       [http://dx.doi.org/10.1177/154405910308201208] [PMID: 14630898]

[50]   Huang GT, Yamaza T, Shea LD, *et al.* Stem/progenitor cell-mediated *de novo* regeneration of dental pulp with newly deposited continuous layer of dentin in an *in vivo* model. Tissue Eng Part A 2010; 16(2): 605-15.
       [http://dx.doi.org/10.1089/ten.tea.2009.0518] [PMID: 19737072]

[51]   El-Backly RM, Massoud AG, El-Badry AM, Sherif RA, Marei MK. Regeneration of dentine/pulp-like tissue using a dental pulp stem cell/poly(lactic-co-glycolic) acid scaffold construct in New Zealand white rabbits. Aust Endod J 2008; 34(2): 52-67.
       [http://dx.doi.org/10.1111/j.1747-4477.2008.00139.x] [PMID: 18666990]

[52]   Ji Y-M, Jeon SH, Park J-Y, Chung J-H, Choung Y-H, Choung P-H. Dental stem cell therapy with calcium hydroxide in dental pulp capping. Tissue Eng Part A 2010; 16(6): 1823-33.
       [http://dx.doi.org/10.1089/ten.tea.2009.0054] [PMID: 20055661]

[53]   Maeda H, Tomokiyo A, Fujii S, Wada N, Akamine A. Promise of periodontal ligament stem cells in regeneration of periodontium. Stem Cell Res Ther 2011; 2(4): 33.
       [http://dx.doi.org/10.1186/scrt74] [PMID: 21861868]

[54]   Zhu W, Liang M. Periodontal ligament stem cells: current status, concerns, and future prospects. Stem Cells Int 2015; 2015: 1-11.

[55]   Li B, Jin Y. Periodontal Tissue Engineering. Stem Cell Biology and Tissue Engineering in Dental Sciences. Elsevier 2015; pp. 471-82.
       [http://dx.doi.org/10.1016/B978-0-12-397157-9.00041-2]

[56]   Seo BM, Song IS, Um S, Lee J-H. Periodontal Ligament Stem Cells. Stem Cell Biology and Tissue Engineering in Dental Sciences. Elsevier 2015; pp. 291-6.

[http://dx.doi.org/10.1016/B978-0-12-397157-9.00024-2]

[57]   Chai Y, Jiang X, Ito Y, *et al*. Fate of the mammalian cranial neural crest during tooth and mandibular morphogenesis. Development 2000; 127(8): 1671-9.
[PMID: 10725243]

[58]   Shin SY, Rios HF, Giannobile WV, Oh T-J. Periodontal Regeneration. Stem Cell Biology and Tissue Engineering in Dental Sciences. Elsevier 2015; pp. 459-69.
[http://dx.doi.org/10.1016/B978-0-12-397157-9.00040-0]

[59]   Rajat Gothi AK. Periodontal Ligament Stem Cells-The Regeneration Front. Dentistry 2015; 5: 275.
[http://dx.doi.org/10.4172/2161-1122.1000275]

[60]   Seo B-M, Miura M, Gronthos S, *et al*. Investigation of multipotent postnatal stem cells from human periodontal ligament. Lancet 2004; 364(9429): 149-55.
[http://dx.doi.org/10.1016/S0140-6736(04)16627-0] [PMID: 15246727]

[61]   Acharya A, Shetty S, Deshmukh V. Periodontal ligament stem cells: An overview. J Oral Biosci 2010; 52(3): 275-82.
[http://dx.doi.org/10.1016/S1349-0079(10)80032-5]

[62]   Kémoun P, Gronthos S, Snead ML, *et al*. The role of cell surface markers and enamel matrix derivatives on human periodontal ligament mesenchymal progenitor responses *in vitro*. Biomaterials 2011; 32(30): 7375-88.
[http://dx.doi.org/10.1016/j.biomaterials.2011.06.043] [PMID: 21784516]

[63]   Wada N, Menicanin D, Shi S, Bartold PM, Gronthos S. Immunomodulatory properties of human periodontal ligament stem cells. J Cell Physiol 2009; 219(3): 667-76.
[http://dx.doi.org/10.1002/jcp.21710] [PMID: 19160415]

[64]   Zhang J, An Y, Gao L-N, Zhang Y-J, Jin Y, Chen F-M. The effect of aging on the pluripotential capacity and regenerative potential of human periodontal ligament stem cells. Biomaterials 2012; 33(29): 6974-86.
[http://dx.doi.org/10.1016/j.biomaterials.2012.06.032] [PMID: 22789721]

[65]   Lekic P, Rojas J, Birek C, Tenenbaum H, McCulloch CA. Phenotypic comparison of periodontal ligament cells *in vivo* and *in vitro*. J Periodontal Res 2001; 36(2): 71-9.
[http://dx.doi.org/10.1034/j.1600-0765.2001.360202.x] [PMID: 11327081]

[66]   Choi S, Cho T-J, Kwon S-K, Lee G, Cho J. Chondrogenesis of periodontal ligament stem cells by transforming growth factor-β3 and bone morphogenetic protein-6 in a normal healthy impacted third molar. Int J Oral Sci 2013; 5(1): 7-13.
[http://dx.doi.org/10.1038/ijos.2013.19] [PMID: 23579467]

[67]   Song D-S, Park J-C, Jung I-H, *et al*. Enhanced adipogenic differentiation and reduced collagen synthesis induced by human periodontal ligament stem cells might underlie the negative effect of recombinant human bone morphogenetic protein-2 on periodontal regeneration. J Periodontal Res 2011; 46(2): 193-203.
[http://dx.doi.org/10.1111/j.1600-0765.2010.01328.x] [PMID: 21118417]

[68]   Fortino VR, Chen R-S, Pelaez D, Cheung HS. Neurogenesis of neural crest-derived periodontal ligament stem cells by EGF and bFGF. J Cell Physiol 2014; 229(4): 479-88.
[http://dx.doi.org/10.1002/jcp.24468] [PMID: 24105823]

[69]   Kawanabe N, Murata S, Murakami K, *et al*. Isolation of multipotent stem cells in human periodontal ligament using stage-specific embryonic antigen-4. Differentiation 2010; 79(2): 74-83.
[http://dx.doi.org/10.1016/j.diff.2009.10.005] [PMID: 19945209]

[70]   Okubo N, Ishisaki A, Iizuka T, Tamura M, Kitagawa Y. Vascular cell-like potential of undifferentiated ligament fibroblasts to construct vascular cell-specific marker-positive blood vessel structures in a PI3K activation-dependent manner. J Vasc Res 2010; 47(5): 369-83.
[http://dx.doi.org/10.1159/000277724] [PMID: 20110728]

[71] Lee JS, An SY, Kwon IK, Heo JS. Transdifferentiation of human periodontal ligament stem cells into pancreatic cell lineage. Cell Biochem Funct 2014; 32(7): 605-11.
[http://dx.doi.org/10.1002/cbf.3057] [PMID: 25187163]

[72] Gay IC, Chen S, MacDougall M. Isolation and characterization of multipotent human periodontal ligament stem cells. Orthod Craniofac Res 2007; 10(3): 149-60.
[http://dx.doi.org/10.1111/j.1601-6343.2007.00399.x] [PMID: 17651131]

[73] Zhang J, Li Z-G, Si Y-M, Chen B, Meng J. The difference on the osteogenic differentiation between periodontal ligament stem cells and bone marrow mesenchymal stem cells under inflammatory microenviroments. Differentiation 2014; 88(4-5): 97-105.
[http://dx.doi.org/10.1016/j.diff.2014.10.001] [PMID: 25498523]

[74] Chen K, Xiong H, Huang Y, Liu C. Comparative analysis of *in vitro* periodontal characteristics of stem cells from apical papilla (SCAP) and periodontal ligament stem cells (PDLSCs). Arch Oral Biol 2013; 58(8): 997-1006.
[http://dx.doi.org/10.1016/j.archoralbio.2013.02.010] [PMID: 23582988]

[75] Pandula PK, Samaranayake LP, Jin LJ, Zhang CF. Human umbilical vein endothelial cells synergize osteo/odontogenic differentiation of periodontal ligament stem cells in 3D cell sheets. J Periodontal Res 2014; 49(3): 299-306.
[http://dx.doi.org/10.1111/jre.12107] [PMID: 23738684]

[76] Zhao L, Wu Y, Tan L, *et al.* Coculture with endothelial cells enhances osteogenic differentiation of periodontal ligament stem cells via cyclooxygenase-2/prostaglandin E_2/vascular endothelial growth factor signaling under hypoxia. J Periodontol 2013; 84(12): 1847-57.
[http://dx.doi.org/10.1902/jop.2013.120548] [PMID: 23537125]

[77] Wu Y, Yang Y, Yang P, *et al.* The osteogenic differentiation of PDLSCs is mediated through MEK/ERK and p38 MAPK signalling under hypoxia. Arch Oral Biol 2013; 58(10): 1357-68.
[http://dx.doi.org/10.1016/j.archoralbio.2013.03.011] [PMID: 23806288]

[78] Tomokiyo A, Maeda H, Fujii S, Wada N, Shima K, Akamine A. Development of a multipotent clonal human periodontal ligament cell line. Differentiation 2008; 76(4): 337-47.
[http://dx.doi.org/10.1111/j.1432-0436.2007.00233.x] [PMID: 18021259]

[79] Moshaverinia A, Xu X, Chen C, *et al.* Application of stem cells derived from the periodontal ligament or gingival tissue sources for tendon tissue regeneration. Biomaterials 2014; 35(9): 2642-50.
[http://dx.doi.org/10.1016/j.biomaterials.2013.12.053] [PMID: 24397989]

[80] Feng F, Akiyama K, Liu Y, *et al.* Utility of PDL progenitors for *in vivo* tissue regeneration: a report of 3 cases. Oral Dis 2010; 16(1): 20-8.
[http://dx.doi.org/10.1111/j.1601-0825.2009.01593.x] [PMID: 20355278]

[81] Liu Y, Zheng Y, Ding G, *et al.* Periodontal ligament stem cell-mediated treatment for periodontitis in miniature swine. Stem Cells 2008; 26(4): 1065-73.
[http://dx.doi.org/10.1634/stemcells.2007-0734] [PMID: 18238856]

[82] Shi S, Bartold PM, Miura M, Seo BM, Robey PG, Gronthos S. The efficacy of mesenchymal stem cells to regenerate and repair dental structures. Orthod Craniofac Res 2005; 8(3): 191-9.
[http://dx.doi.org/10.1111/j.1601-6343.2005.00331.x] [PMID: 16022721]

[83] Sonoyama W, Liu Y, Fang D, *et al.* Mesenchymal Stem Cell-Mediated Functional Tooth Regeneration in Swine. 2006.
[http://dx.doi.org/10.1371/journal.pone.0000079]

[84] Yang Z, Jin F, Zhang X, *et al.* Tissue engineering of cementum/periodontal-ligament complex using a novel three-dimensional pellet cultivation system for human periodontal ligament stem cells. Tissue Eng Part C Methods 2009; 15(4): 571-81.
[http://dx.doi.org/10.1089/ten.tec.2008.0561] [PMID: 19534606]

[85] Flores MG, Hasegawa M, Yamato M, Takagi R, Okano T, Ishikawa I. Cementum-periodontal

ligament complex regeneration using the cell sheet technique. J Periodontal Res 2008; 43(3): 364-71.
[http://dx.doi.org/10.1111/j.1600-0765.2007.01046.x] [PMID: 18205734]

[86]   Zhou Y, Li Y, Mao L, Peng H. Periodontal healing by periodontal ligament cell sheets in a teeth
       replantation model. Arch Oral Biol 2012; 57(2): 169-76.
       [http://dx.doi.org/10.1016/j.archoralbio.2011.08.008] [PMID: 21907971]

[87]   Akizuki T, Oda S, Komaki M, *et al.* Application of periodontal ligament cell sheet for periodontal
       regeneration: a pilot study in beagle dogs. J Periodontal Res 2005; 40(3): 245-51.
       [http://dx.doi.org/10.1111/j.1600-0765.2005.00799.x] [PMID: 15853971]

[88]   Hosoya A, Ninomiya T, Hiraga T, *et al.* Potential of Periodontal Ligament Cells to Regenerate
       Alveolar Bone. J Oral Biosci 2010; 52(2): 72-80.
       [http://dx.doi.org/10.1016/S1349-0079(10)80035-0]

[89]   Seo B-M, Miura M, Sonoyama W, Coppe C, Stanyon R, Shi S. Recovery of stem cells from
       cryopreserved periodontal ligament. J Dent Res 2005; 84(10): 907-12.
       [http://dx.doi.org/10.1177/154405910508401007] [PMID: 16183789]

[90]   Vasconcelos RG, Ribeiro RA, Vasconcelos MG, Lima KC, Barboza CA. *In vitro* comparative analysis
       of cryopreservation of undifferentiated mesenchymal cells derived from human periodontal ligament.
       Cell Tissue Bank 2012; 13(3): 461-9.
       [http://dx.doi.org/10.1007/s10561-011-9271-3] [PMID: 21833489]

[91]   Ding G, Liu Y, Wang W, *et al.* Allogeneic periodontal ligament stem cell therapy for periodontitis in
       swine. Stem Cells 2010; 28(10): 1829-38.
       [http://dx.doi.org/10.1002/stem.512] [PMID: 20979138]

[92]   Ericson S, Bjerklin K. The dental follicle in normally and ectopically erupting maxillary canines: a
       computed tomography study. Angle Orthod 2001; 71(5): 333-42.
       [PMID: 11605866]

[93]   Honda MJ, Tsuchiya S, Shinohara Y, Shinmura Y, Sumita Y. Recent advances in engineering of tooth
       and tooth structures using postnatal dental cells. Jpn Dent Sci Rev 2010; 46(1): 54-66.
       [http://dx.doi.org/10.1016/j.jdsr.2009.10.006]

[94]   Morsczeck C. Dental Follicle Stem Cells. Stem Cell Biology and Tissue Engineering in Dental
       Sciences. Elsevier 2015; pp. 271-7.
       [http://dx.doi.org/10.1016/B978-0-12-397157-9.00022-9]

[95]   Wise GE. Cellular and molecular basis of tooth eruption. Orthod Craniofac Res 2009; 12(2): 67-73.
       [http://dx.doi.org/10.1111/j.1601-6343.2009.01439.x] [PMID: 19419449]

[96]   Morsczeck C, Götz W, Schierholz J, *et al.* Isolation of precursor cells (PCs) from human dental follicle
       of wisdom teeth. Matrix Biol 2005; 24(2): 155-65.
       [http://dx.doi.org/10.1016/j.matbio.2004.12.004] [PMID: 15890265]

[97]   Han C, Yang Z, Zhou W, *et al.* Periapical follicle stem cell: a promising candidate for
       cementum/periodontal ligament regeneration and bio-root engineering. Stem Cells Dev 2010; 19(9):
       1405-15.
       [http://dx.doi.org/10.1089/scd.2009.0277] [PMID: 19995154]

[98]   Li X, Yang C, Li L, *et al.* A therapeutic strategy for spinal cord defect: human dental follicle cells
       combined with aligned PCL/PLGA electrospun material. BioMed Res Int 2015; 2015: 197183.
       [PMID: 25695050]

[99]   Navabazam AR, Sadeghian Nodoshan F, Sheikhha MH, Miresmaeili SM, Soleimani M, Fesahat F.
       Characterization of mesenchymal stem cells from human dental pulp, preapical follicle and
       periodontal ligament. Iran J Reprod Med 2013; 11(3): 235-42.
       [PMID: 24639751]

[100] Shoi K, Aoki K, Ohya K, Takagi Y, Shimokawa H. Characterization of pulp and follicle stem cells
       from impacted supernumerary maxillary incisors. Pediatr Dent 2014; 36(3): 79-84.

[PMID: 24960375]

[101] Kang Y-H, Lee H-J, Jang S-J, *et al.* Immunomodulatory properties and *in vivo* osteogenesis of human
       dental stem cells from fresh and cryopreserved dental follicles. Differentiation 2015; 90(1-3): 48-58.
       [http://dx.doi.org/10.1016/j.diff.2015.10.001] [PMID: 26493125]

[102] Yildirim S, Zibandeh N, Genc D, Ozcan EM, Goker K, Akkoc T. The comparison of the immunologic
       properties of stem cells isolated from human exfoliated deciduous teeth, dental pulp, and dental
       follicles. Stem Cells Int 2016; 2016: 4682875.

[103] Mori G, Ballini A, Carbone C, *et al.* Osteogenic differentiation of dental follicle stem cells. Int J Med
       Sci 2012; 9(6): 480-7.
       [http://dx.doi.org/10.7150/ijms.4583] [PMID: 22927773]

[104] Felthaus O, Ernst W, Driemel O, Reichert TE, Schmalz G, Morsczeck C. TGF-β stimulates glial-like
       differentiation in murine dental follicle precursor cells (mDFPCs). Neurosci Lett 2010; 471(3): 179-
       84.
       [http://dx.doi.org/10.1016/j.neulet.2010.01.037] [PMID: 20100544]

[105] Morsczeck C, Völlner F, Saugspier M, *et al.* Comparison of human dental follicle cells (DFCs) and
       stem cells from human exfoliated deciduous teeth (SHED) after neural differentiation *in vitro.* Clin
       Oral Investig 2010; 14(4): 433-40.
       [http://dx.doi.org/10.1007/s00784-009-0310-4] [PMID: 19590907]

[106] Völlner F, Ernst W, Driemel O, Morsczeck C. A two-step strategy for neuronal differentiation *in vitro*
       of human dental follicle cells. Differentiation 2009; 77(5): 433-41.
       [http://dx.doi.org/10.1016/j.diff.2009.03.002] [PMID: 19394129]

[107] Yao S, Pan F, Prpic V, Wise GE. Differentiation of stem cells in the dental follicle. J Dent Res 2008;
       87(8): 767-71.
       [http://dx.doi.org/10.1177/154405910808700801] [PMID: 18650550]

[108] Park B-W, Kang E-J, Byun J-H, *et al. In vitro* and *in vivo* osteogenesis of human mesenchymal stem
       cells derived from skin, bone marrow and dental follicle tissues. Differentiation 2012; 83(5): 249-59.
       [http://dx.doi.org/10.1016/j.diff.2012.02.008] [PMID: 22469856]

[109] Takahashi K, Ogura N, Tomoki R, *et al.* Applicability of human dental follicle cells to bone
       regeneration without dexamethasone: an *in vivo* pilot study. Int J Oral Maxillofac Surg 2015; 44(5):
       664-9.
       [http://dx.doi.org/10.1016/j.ijom.2014.11.006] [PMID: 25496849]

[110] Rezai-Rad M, Bova JF, Orooji M, *et al.* Evaluation of bone regeneration potential of dental follicle
       stem cells for treatment of craniofacial defects. Cytotherapy 2015; 17(11): 1572-81.
       [http://dx.doi.org/10.1016/j.jcyt.2015.07.013] [PMID: 26342992]

[111] Chen G, Sun Q, Xie L, *et al.* Comparison of the Odontogenic Differentiation Potential of Dental
       Follicle, Dental Papilla, and Cranial Neural Crest Cells. J Endod 2015; 41(7): 1091-9.
       [http://dx.doi.org/10.1016/j.joen.2015.03.003] [PMID: 25882137]

[112] Guo W, He Y, Zhang X, *et al.* The use of dentin matrix scaffold and dental follicle cells for dentin
       regeneration. Biomaterials 2009; 30(35): 6708-23.
       [http://dx.doi.org/10.1016/j.biomaterials.2009.08.034] [PMID: 19767098]

[113] Yokoi T, Saito M, Kiyono T, *et al.* Establishment of immortalized dental follicle cells for generating
       periodontal ligament *in vivo.* Cell Tissue Res 2007; 327(2): 301-11.
       [http://dx.doi.org/10.1007/s00441-006-0257-6] [PMID: 17013589]

[114] Bai Y, Bai Y, Matsuzaka K, *et al.* Cementum- and periodontal ligament-like tissue formation by dental
       follicle cell sheets co-cultured with Hertwig's epithelial root sheath cells. Bone 2011; 48(6): 1417-26.
       [http://dx.doi.org/10.1016/j.bone.2011.02.016] [PMID: 21376148]

[115] Sowmya S, Chennazhi KP, Arzate H, Jayachandran P, Nair SV, Jayakumar R. Periodontal Specific
       Differentiation of Dental Follicle Stem Cells into Osteoblast, Fibroblast, and Cementoblast. Tissue

Eng Part C Methods 2015; 21(10): 1044-58.
[http://dx.doi.org/10.1089/ten.tec.2014.0603] [PMID: 25962715]

[116]   Yang B, Chen G, Li J, *et al.* Tooth root regeneration using dental follicle cell sheets in combination with a dentin matrix - based scaffold. Biomaterials 2012; 33(8): 2449-61.
[http://dx.doi.org/10.1016/j.biomaterials.2011.11.074] [PMID: 22192537]

[117]   Wu J, Li W, Dai J, Duan Y, Jin Y. Dentin elasticity may contribute to the differentiation of dental follicle cells into cementoblast lineages. Biosci Hypotheses 2008; 1(1): 2-4.
[http://dx.doi.org/10.1016/j.bihy.2008.01.005]

[118]   Wei X, Yang X, Han ZP, Qu FF, Shao L, Shi YF. Mesenchymal stem cells: a new trend for cell therapy. Acta Pharmacol Sin 2013; 34(6): 747-54.
[http://dx.doi.org/10.1038/aps.2013.50] [PMID: 23736003]

[119]   Ullah I, Subbarao RB, Rho GJ. Human mesenchymal stem cells - current trends and future prospective. Biosci Rep 2015; 35(2): 1-18.
[http://dx.doi.org/10.1042/BSR20150025] [PMID: 25797907]

[120]   Egusa H, Sonoyama W, Nishimura M, Atsuta I, Akiyama K. Stem cells in dentistry--part I: stem cell sources. J Prosthodont Res 2012; 56(3): 151-65.
[http://dx.doi.org/10.1016/j.jpor.2012.06.001] [PMID: 22796367]

[121]   Lee B-K, Choi S-J, Mack D, Oh S-H. Isolation of mesenchymal stem cells from the mandibular marrow aspirates. Oral Surg Oral Med Oral Pathol Oral Radiol Endod 2011; 112(6): e86-93.
[http://dx.doi.org/10.1016/j.tripleo.2011.05.032] [PMID: 21872505]

[122]   Hosoya A, Ninomiya T, Hiraga T, *et al.* Alveolar bone regeneration of subcutaneously transplanted rat molar. Bone 2008; 42(2): 350-7.
[http://dx.doi.org/10.1016/j.bone.2007.09.054] [PMID: 18032126]

[123]   Mao JJ, Prockop DJ. Stem cells in the face: tooth regeneration and beyond. Cell Stem Cell 2012; 11(3): 291-301.
[http://dx.doi.org/10.1016/j.stem.2012.08.010] [PMID: 22958928]

[124]   Aghaloo TL, Chaichanasakul T, Bezouglaia O, *et al.* Osteogenic potential of mandibular *vs.* long-bone marrow stromal cells. J Dent Res 2010; 89(11): 1293-8.
[http://dx.doi.org/10.1177/0022034510378427] [PMID: 20811069]

[125]   Akintoye SO, Lam T, Shi S, Brahim J, Collins MT, Robey PG. Skeletal site-specific characterization of orofacial and iliac crest human bone marrow stromal cells in same individuals. Bone 2006; 38(6): 758-68.
[http://dx.doi.org/10.1016/j.bone.2005.10.027] [PMID: 16403496]

[126]   Igarashi A, Segoshi K, Sakai Y, *et al.* Selection of common markers for bone marrow stromal cells from various bones using real-time RT-PCR: effects of passage number and donor age. Tissue Eng 2007; 13(10): 2405-17.
[http://dx.doi.org/10.1089/ten.2006.0340] [PMID: 17596118]

[127]   Nishimura M, Takase K, Suehiro F, Murata H. Candidates cell sources to regenerate alveolar bone from oral tissue. Int J Dent 2012; 2012: 857192.
[http://dx.doi.org/10.1155/2012/857192]

[128]   de Souza Faloni AP, Schoenmaker T, Azari A, *et al.* Jaw and long bone marrows have a different osteoclastogenic potential. Calcif Tissue Int 2011; 88(1): 63-74.
[http://dx.doi.org/10.1007/s00223-010-9418-4] [PMID: 20862464]

[129]   Osyczka AM, Damek-Poprawa M, Wojtowicz A, Akintoye SO. Age and skeletal sites affect BMP-2 responsiveness of human bone marrow stromal cells. Connect Tissue Res 2009; 50(4): 270-7.
[http://dx.doi.org/10.1080/03008200902846262] [PMID: 19637063]

[130]   Akintoye SO, Giavis P, Stefanik D, Levin L, Mante FK. Comparative osteogenesis of maxilla and iliac crest human bone marrow stromal cells attached to oxidized titanium: a pilot study. Clin Oral Implants

Res 2008; 19(11): 1197-201.
[http://dx.doi.org/10.1111/j.1600-0501.2008.01592.x] [PMID: 18983324]

[131] Yamaza T, Ren G, Akiyama K, Chen C, Shi Y, Shi S. Mouse mandible contains distinctive mesenchymal stem cells. J Dent Res 2011; 90(3): 317-24.
[http://dx.doi.org/10.1177/0022034510387796] [PMID: 21076121]

[132] Steinhardt Y, Aslan H, Regev E, *et al.* Maxillofacial-derived stem cells regenerate critical mandibular bone defect. Tissue Eng Part A 2008; 14(11): 1763-73.
[http://dx.doi.org/10.1089/ten.tea.2008.0007] [PMID: 18636943]

[133] Marolt D, Rode M, Kregar-Velikonja N, Jeras M, Knezevic M. Primary human alveolar bone cells isolated from tissue samples acquired at periodontal surgeries exhibit sustained proliferation and retain osteogenic phenotype during *in vitro* expansion. In: Gerecht S, Ed. PLoS ONE 2014; 9(3): e92969.

[134] Matsubara T, Suardita K, Ishii M, *et al.* Alveolar bone marrow as a cell source for regenerative medicine: differences between alveolar and iliac bone marrow stromal cells. J Bone Miner Res 2005; 20(3): 399-409.
[http://dx.doi.org/10.1359/JBMR.041117] [PMID: 15746984]

[135] Damek-Poprawa M, Stefanik D, Levin LM, Akintoye SO. Human bone marrow stromal cells display variable anatomic site-dependent response and recovery from irradiation. Arch Oral Biol 2010; 55(5): 358-64.
[http://dx.doi.org/10.1016/j.archoralbio.2010.03.010] [PMID: 20378097]

[136] Keller L, Offner D, Schwinté P, *et al.* Active Nanomaterials to Meet the Challenge of Dental Pulp Regeneration. Materials (Basel) 2015; 8(11): 7461-71.
[http://dx.doi.org/10.3390/ma8115387]

[137] Huang GT. Dental pulp and dentin tissue engineering and regeneration: advancement and challenge. Front Biosci (Elite Ed) 2011; 3: 788-800.
[http://dx.doi.org/10.2741/e286] [PMID: 21196351]

[138] Ravindran S, George A. Biomimetic extracellular matrix mediated somatic stem cell differentiation: applications in dental pulp tissue regeneration. Front Physiol 2015; 6: 118.
[http://dx.doi.org/10.3389/fphys.2015.00118] [PMID: 25954205]

[139] Zhang W, Yelick PC. Vital pulp therapy—current progress of dental pulp regeneration and revascularization. Int J Dent 2010; 2010: 856087.

[140] Kim S, Shin S-J, Song Y, Kim E. *In vivo* experiments with dental pulp stem cells for pulp-dentin complex regeneration. Mediators Inflamm 2015; 2015: 409347.

[141] Nakashima M, Iohara K. Mobilized dental pulp stem cells for pulp regeneration: initiation of clinical trial. J Endod 2014; 40(4) (Suppl.): S26-32.
[http://dx.doi.org/10.1016/j.joen.2014.01.020] [PMID: 24698690]

[142] Wang Y, Zhao Y, Jia W, Yang J, Ge L. Preliminary study on dental pulp stem cell-mediated pulp regeneration in canine immature permanent teeth. J Endod 2013; 39(2): 195-201.
[http://dx.doi.org/10.1016/j.joen.2012.10.002] [PMID: 23321230]

[143] Alongi DJ, Yamaza T, Song Y, *et al.* Stem/progenitor cells from inflamed human dental pulp retain tissue regeneration potential. Regen Med 2010; 5(4): 617-31.
[http://dx.doi.org/10.2217/rme.10.30] [PMID: 20465527]

[144] Xu W, Jiang S, Chen Q, *et al.* Systemically Transplanted Bone Marrow-derived Cells Contribute to Dental Pulp Regeneration in a Chimeric Mouse Model. J Endod 2016; 42(2): 263-8.
[http://dx.doi.org/10.1016/j.joen.2015.10.007] [PMID: 26686823]

[145] Zhang L-X, Shen L-L, Ge S-H, *et al.* Systemic BMSC homing in the regeneration of pulp-like tissue and the enhancing effect of stromal cell-derived factor-1 on BMSC homing. Int J Clin Exp Pathol 2015; 8(9): 10261-71.
[PMID: 26617734]

[146] Ishizaka R, Iohara K, Murakami M, Fukuta O, Nakashima M. Regeneration of dental pulp following pulpectomy by fractionated stem/progenitor cells from bone marrow and adipose tissue. Biomaterials 2012; 33(7): 2109-18.
[http://dx.doi.org/10.1016/j.biomaterials.2011.11.056] [PMID: 22177838]

[147] Prescott RS, Alsanea R, Fayad MI, *et al*. *In vivo* generation of dental pulp-like tissue by using dental pulp stem cells, a collagen scaffold, and dentin matrix protein 1 after subcutaneous transplantation in mice. J Endod 2008; 34(4): 421-6.
[http://dx.doi.org/10.1016/j.joen.2008.02.005] [PMID: 18358888]

[148] Srisuwan T, Tilkorn DJ, Al-Benna S, Abberton K, Messer HH, Thompson EW. Revascularization and tissue regeneration of an empty root canal space is enhanced by a direct blood supply and stem cells. Dent Traumatol 2013; 29(2): 84-91.
[http://dx.doi.org/10.1111/j.1600-9657.2012.01136.x] [PMID: 22520279]

[149] Ravindran S, Zhang Y, Huang C-C, George A. Odontogenic induction of dental stem cells by extracellular matrix-inspired three-dimensional scaffold. Tissue Eng Part A 2014; 20(1-2): 92-102.
[http://dx.doi.org/10.1089/ten.tea.2013.0192] [PMID: 23859633]

[150] Erisken C, Kalyon DM, Zhou J, Kim SG, Mao JJ. Viscoelastic Properties of Dental Pulp Tissue and Ramifications on Biomaterial Development for Pulp Regeneration. J Endod 2015; 41(10): 1711-7.
[http://dx.doi.org/10.1016/j.joen.2015.07.005] [PMID: 26321063]

[151] Ferroni L, Gardin C, Sivolella S, *et al*. A hyaluronan-based scaffold for the *in vitro* construction of dental pulp-like tissue. Int J Mol Sci 2015; 16(3): 4666-81.
[http://dx.doi.org/10.3390/ijms16034666] [PMID: 25739081]

[152] Coimbra P, Alves P, Valente TA, Santos R, Correia IJ, Ferreira P. Sodium hyaluronate/chitosan polyelectrolyte complex scaffolds for dental pulp regeneration: synthesis and characterization. Int J Biol Macromol 2011; 49(4): 573-9.
[http://dx.doi.org/10.1016/j.ijbiomac.2011.06.011] [PMID: 21704650]

[153] Yang JW, Zhang YF, Sun ZY, Song GT, Chen Z. Dental pulp tissue engineering with bFGF-incorporated silk fibroin scaffolds. J Biomater Appl 2015; 30(2): 221-9.
[http://dx.doi.org/10.1177/0885328215577296] [PMID: 25791684]

[154] Cordeiro MM, Dong Z, Kaneko T, *et al*. Dental pulp tissue engineering with stem cells from exfoliated deciduous teeth. J Endod 2008; 34(8): 962-9.
[http://dx.doi.org/10.1016/j.joen.2008.04.009] [PMID: 18634928]

[155] Chen G, Chen J, Yang B, *et al*. Combination of aligned PLGA/Gelatin electrospun sheets, native dental pulp extracellular matrix and treated dentin matrix as substrates for tooth root regeneration. Biomaterials 2015; 52: 56-70.
[http://dx.doi.org/10.1016/j.biomaterials.2015.02.011] [PMID: 25818413]

[156] Casagrande L, Demarco FF, Zhang Z, Araujo FB, Shi S, Nör JE. Dentin-derived BMP-2 and odontoblast differentiation. J Dent Res 2010; 89(6): 603-8.
[http://dx.doi.org/10.1177/0022034510364487] [PMID: 20351355]

[157] Rutherford RB, Spångberg L, Tucker M, Rueger D, Charette M. The time-course of the induction of reparative dentine formation in monkeys by recombinant human osteogenic protein-1. Arch Oral Biol 1994; 39(10): 833-8.
[http://dx.doi.org/10.1016/0003-9969(94)90014-0] [PMID: 7741652]

[158] Six N, Decup F, Lasfargues J-J, Salih E, Goldberg M. Osteogenic proteins (bone sialoprotein and bone morphogenetic protein-7) and dental pulp mineralization. J Mater Sci Mater Med 2002; 13(2): 225-32.
[http://dx.doi.org/10.1023/A:1013846516693] [PMID: 15348647]

[159] Decup F, Six N, Palmier B, *et al*. Bone sialoprotein-induced reparative dentinogenesis in the pulp of rat's molar. Clin Oral Investig 2000; 4(2): 110-9.
[http://dx.doi.org/10.1007/s007840050126] [PMID: 11218498]

[160] Iohara K, Murakami M, Takeuchi N, *et al.* A novel combinatorial therapy with pulp stem cells and granulocyte colony-stimulating factor for total pulp regeneration. Stem Cells Transl Med 2013; 2(7): 521-33.
[http://dx.doi.org/10.5966/sctm.2012-0132] [PMID: 23761108]

[161] Lovschall H, Fejerskov O, Flyvbjerg A. Pulp-capping with recombinant human insulin-like growth factor I (rhIGF-I) in rat molars. Adv Dent Res 2001; 15: 108-12.
[http://dx.doi.org/10.1177/08959374010150010301] [PMID: 12640754]

[162] Morito A, Kida Y, Suzuki K, *et al.* Effects of basic fibroblast growth factor on the development of the stem cell properties of human dental pulp cells. Arch Histol Cytol 2009; 72(1): 51-64.
[http://dx.doi.org/10.1679/aohc.72.51] [PMID: 19789412]

[163] Suzuki T, Lee CH, Chen M, *et al.* Induced migration of dental pulp stem cells for *in vivo* pulp regeneration. J Dent Res 2011; 90(8): 1013-8.
[http://dx.doi.org/10.1177/0022034511408426] [PMID: 21586666]

[164] Kim JY, Xin X, Moioli EK, *et al.* Regeneration of dental-pulp-like tissue by chemotaxis-induced cell homing. Tissue Eng Part A 2010; 16(10): 3023-31.
[http://dx.doi.org/10.1089/ten.tea.2010.0181] [PMID: 20486799]

[165] Kitamura C, Nishihara T, Terashita M, Tabata Y, Washio A. Local regeneration of dentin-pulp complex using controlled release of fgf-2 and naturally derived sponge-like scaffolds. Int J Dent 2012; 2012: 190561.

[166] Requicha JF, Viegas CA, Muñoz F, Reis RL, Gomes ME. Periodontal tissue engineering strategies based on nonoral stem cells. Anat Rec (Hoboken) 2014; 297(1): 6-15.
[http://dx.doi.org/10.1002/ar.22797] [PMID: 24293355]

[167] Dan H, Vaquette C, Fisher AG, *et al.* The influence of cellular source on periodontal regeneration using calcium phosphate coated polycaprolactone scaffold supported cell sheets. Biomaterials 2014; 35(1): 113-22.
[http://dx.doi.org/10.1016/j.biomaterials.2013.09.074] [PMID: 24120045]

[168] Chen F-M, Sun H-H, Lu H, Yu Q. Stem cell-delivery therapeutics for periodontal tissue regeneration. Biomaterials 2012; 33(27): 6320-44.
[http://dx.doi.org/10.1016/j.biomaterials.2012.05.048] [PMID: 22695066]

[169] Chen F-M, Shi S. Periodontal Tissue Engineering. Principles of Tissue Engineering. Elsevier 2014; pp. 1507-40.
[http://dx.doi.org/10.1016/B978-0-12-398358-9.00072-0]

[170] Yang Y, Rossi FM, Putnins EE. Periodontal regeneration using engineered bone marrow mesenchymal stromal cells. Biomaterials 2010; 31(33): 8574-82.
[http://dx.doi.org/10.1016/j.biomaterials.2010.06.026] [PMID: 20832109]

[171] Sowmya S, Bumgardener JD, Chennazhi KP, Nair SV, Jayakumar R. Role of nanostructured biopolymers and bioceramics in enamel, dentin and periodontal tissue regeneration. Prog Polym Sci 2013; 38(10-11): 1748-72.
[http://dx.doi.org/10.1016/j.progpolymsci.2013.05.005]

[172] Guo W, Chen L, Gong K, Ding B, Duan Y, Jin Y. Heterogeneous dental follicle cells and the regeneration of complex periodontal tissues. Tissue Eng Part A 2012; 18(5-6): 459-70.
[http://dx.doi.org/10.1089/ten.tea.2011.0261] [PMID: 21919800]

[173] Takedachi M, Sawada K, Yamamoto S, *et al.* Periodontal tissue regeneration by transplantation of adipose tissue-derived stem cells. J Oral Biosci 2013; 55(3): 137-42.
[http://dx.doi.org/10.1016/j.job.2013.04.004]

[174] Tobita M, Uysal AC, Ogawa R, Hyakusoku H, Mizuno H. Periodontal tissue regeneration with adipose-derived stem cells. Tissue Eng Part A 2008; 14(6): 945-53.
[http://dx.doi.org/10.1089/ten.tea.2007.0048] [PMID: 18558814]

[175]   Kawaguchi H, Hirachi A, Hasegawa N, *et al.* Enhancement of periodontal tissue regeneration by transplantation of bone marrow mesenchymal stem cells. J Periodontol 2004; 75(9): 1281-7.
[http://dx.doi.org/10.1902/jop.2004.75.9.1281] [PMID: 15515346]

[176]   Inanç B, Elçin AE, Elçin YM. *In vitro* differentiation and attachment of human embryonic stem cells on periodontal tooth root surfaces. Tissue Eng Part A 2009; 15(11): 3427-35.
[http://dx.doi.org/10.1089/ten.tea.2008.0380] [PMID: 19405785]

[177]   Shimauchi H, Nemoto E, Ishihata H, Shimomura M. Possible functional scaffolds for periodontal regeneration. Jpn Dent Sci Rev 2013; 49(4): 118-30.
[http://dx.doi.org/10.1016/j.jdsr.2013.05.001]

[178]   Su F, Liu S-S, Ma J-L, Wang D-S, e LL, Liu HC. Enhancement of periodontal tissue regeneration by transplantation of osteoprotegerin-engineered periodontal ligament stem cells. Stem Cell Res Ther 2015; 6(1): 22.
[http://dx.doi.org/10.1186/s13287-015-0023-3] [PMID: 25888745]

[179]   Park CH, Rios HF, Jin Q, *et al.* Tissue engineering bone-ligament complexes using fiber-guiding scaffolds. Biomaterials 2012; 33(1): 137-45.
[http://dx.doi.org/10.1016/j.biomaterials.2011.09.057] [PMID: 21993234]

[180]   Zhou W, Mei L. Effect of autologous bone marrow stromal cells transduced with osteoprotegerin on periodontal bone regeneration in canine periodontal window defects. Int J Periodontics Restorative Dent 2012; 32(5): e174-81.
[PMID: 22754911]

[181]   Wang Y, Xia H, Zhao Y, Jiang T. Three-dimensional culture of human periodontal ligament cells on highly porous polyglycolic acid scaffolds *in vitro*. Conf Proc Annu Int Conf IEEE Eng Med Biol Soc IEEE Eng Med Biol Soc Annu Conf. 4908-11.

[182]   Gault P, Black A, Romette J-L, *et al.* Tissue-engineered ligament: implant constructs for tooth replacement. J Clin Periodontol 2010; 37(8): 750-8.
[PMID: 20546087]

[183]   Nagayasu-Tanaka T, Anzai J, Takaki S, *et al.* Action mechanism of fibroblast growth factor-2 (fgf-2) in the promotion of periodontal regeneration in beagle dogs. In: Matsumoto T, Ed. PLOS ONE. 2015; 10: p. (6)e0131870.

[184]   Zhang R, Zhang M, Li CH, Wang PC, Chen F, Wang QT. Effects of basic fibroblast growth factor and vascular endothelial growth factor on the proliferation, migration and adhesion of human periodontal ligament stem cells *in vitro*. Zhonghua Kou Qiang Yi Xue Za Zhi 2013; 48(5): 278-84.
[PMID: 24004623]

[185]   Murakami S, Takayama S, Kitamura M, *et al.* Recombinant human basic fibroblast growth factor (bFGF) stimulates periodontal regeneration in class II furcation defects created in beagle dogs. J Periodontal Res 2003; 38(1): 97-103.
[http://dx.doi.org/10.1034/j.1600-0765.2003.00640.x] [PMID: 12558943]

[186]   Kitamura M, Akamatsu M, Machigashira M, *et al.* FGF-2 stimulates periodontal regeneration: results of a multi-center randomized clinical trial. J Dent Res 2011; 90(1): 35-40.
[http://dx.doi.org/10.1177/0022034510384616] [PMID: 21059869]

[187]   Takayama S, Murakami S, Shimabukuro Y, Kitamura M, Okada H. Periodontal regeneration by FGF-2 (bFGF) in primate models. J Dent Res 2001; 80(12): 2075-9.
[http://dx.doi.org/10.1177/00220345010800121001] [PMID: 11808765]

[188]   Fujii S, Maeda H, Tomokiyo A, *et al.* Effects of TGF-β1 on the proliferation and differentiation of human periodontal ligament cells and a human periodontal ligament stem/progenitor cell line. Cell Tissue Res 2010; 342(2): 233-42.
[http://dx.doi.org/10.1007/s00441-010-1037-x] [PMID: 20931341]

[189]   Fujita T, Shiba H, Van Dyke TE, Kurihara H. Differential effects of growth factors and cytokines on

the synthesis of SPARC, DNA, fibronectin and alkaline phosphatase activity in human periodontal ligament cells. Cell Biol Int 2004; 28(4): 281-6.
[http://dx.doi.org/10.1016/j.cellbi.2003.12.007] [PMID: 15109984]

[190] Li Y-F, Yan F-H, Zhong Q, Zhao X. Effect of hBMP-7 gene modified bone marrow stromal cells on periodontal tissue regeneration. Zhonghua Yi Xue Za Zhi 2010; 90(20): 1427-30.
[PMID: 20646636]

[191] Pilipchuk SP, Plonka AB, Monje A, *et al.* Tissue engineering for bone regeneration and osseointegration in the oral cavity. Dent Mater 2015; 31(4): 317-38.
[http://dx.doi.org/10.1016/j.dental.2015.01.006] [PMID: 25701146]

[192] Rastogi P, Saini H, Singhal R, Dixit J. Periodontal regeneration in deep intrabony periodontal defect using hydroxyapatite particles with platelet rich fibrin membrane-a case report. J Oral Biol Craniofac Res 2011; 1(1): 41-3.
[http://dx.doi.org/10.1016/S2212-4268(11)60010-9] [PMID: 25756017]

[193] Lang NP. Focus on intrabony defects--conservative therapy. Periodontol 2000 2000; 22: 51-8.
[http://dx.doi.org/10.1034/j.1600-0757.2000.2220105.x] [PMID: 11276516]

[194] Iwata T, Yamato M, Ishikawa I, Ando T, Okano T. Tissue engineering in periodontal tissue. Anat Rec (Hoboken) 2014; 297(1): 16-25.
[http://dx.doi.org/10.1002/ar.22812] [PMID: 24343910]

[195] Jahanbin A, Rashed R, Alamdari DH, *et al.* Success of Maxillary Alveolar Defect Repair in Rats Using Osteoblast-Differentiated Human Deciduous Dental Pulp Stem Cells. J Oral Maxillofac Surg 2016; 74(4): 829.e1-9.
[http://dx.doi.org/10.1016/j.joms.2015.11.033] [PMID: 26763080]

[196] Rai R. Ranjana, Mohan, D, Arunagiri, Vandana, Sharma, Zoya, Chowdhary. Stem Cells in the Regenerative Management of Intra-bony Defects: A Clinical Study. J Dental Coll Azamgarh 2015; 1(1): 44-9.

[197] Tsumanuma Y, Iwata T, Washio K, *et al.* Comparison of different tissue-derived stem cell sheets for periodontal regeneration in a canine 1-wall defect model. Biomaterials 2011; 32(25): 5819-25.
[http://dx.doi.org/10.1016/j.biomaterials.2011.04.071] [PMID: 21605900]

[198] Jain R, Kaur H, Jain S, Kapoor D, Nanda T, Jain M. Comparison of nano-sized hydroxyapatite and β-tricalcium phosphate in the treatment of human periodontal intrabony defects. J Clin Diagn Res 2014; 8(10): ZC74-8.
[PMID: 25478453]

[199] Vaquette C, Fan W, Xiao Y, Hamlet S, Hutmacher DW, Ivanovski S. A biphasic scaffold design combined with cell sheet technology for simultaneous regeneration of alveolar bone/periodontal ligament complex. Biomaterials 2012; 33(22): 5560-73.
[http://dx.doi.org/10.1016/j.biomaterials.2012.04.038] [PMID: 22575832]

[200] Jin Q-M, Zhao M, Webb SA, Berry JE, Somerman MJ, Giannobile WV. Cementum engineering with three-dimensional polymer scaffolds. J Biomed Mater Res A 2003; 67(1): 54-60.
[http://dx.doi.org/10.1002/jbm.a.10058] [PMID: 14517861]

[201] Vandana KL, Desai R, Dalvi PJ. Autologous stem cell application in periodontal regeneration technique (SAI-PRT) using PDLSCs directly from an extracted tooth···an insight. Int J Stem Cells 2015; 8(2): 235-7.
[http://dx.doi.org/10.15283/ijsc.2015.8.2.235] [PMID: 26634072]

[202] Marolt D. Tissue engineering craniofacial bone products. Stem Cell Biology and Tissue Engineering in Dental Sciences. Elsevier 2015; pp. 521-39.
[http://dx.doi.org/10.1016/B978-0-12-397157-9.00044-8]

[203] Nevins ML, Camelo M, Rebaudi A, Lynch SE, Nevins M. Three-dimensional micro-computed tomographic evaluation of periodontal regeneration: a human report of intrabony defects treated with

Bio-Oss collagen. Int J Periodontics Restorative Dent 2005; 25(4): 365-73.
[PMID: 16089044]

[204] Neamat A, Gawish A, Gamal-Eldeen AM. β-Tricalcium phosphate promotes cell proliferation, osteogenesis and bone regeneration in intrabony defects in dogs. Arch Oral Biol 2009; 54(12): 1083-90.
[http://dx.doi.org/10.1016/j.archoralbio.2009.09.003] [PMID: 19828137]

[205] Yukna RA, Yukna CN. A 5-year follow-up of 16 patients treated with coralline calcium carbonate (BIOCORAL) bone replacement grafts in infrabony defects. J Clin Periodontol 1998; 25(12): 1036-40.
[http://dx.doi.org/10.1111/j.1600-051X.1998.tb02410.x] [PMID: 9869355]

[206] Panda S, Jayakumar ND, Sankari M, Varghese SS, Mehta P. Platelet rich fibrin and alloplast in treatment of intrabony defect. J Pharm Res 2013; 7(7): 621-5.
[http://dx.doi.org/10.1016/j.jopr.2013.07.023]

[207] Su N-Y, Chang Y-C. Clinical application of platelet-rich fibrin in perio-endo combined intrabony defect. J Dent Sci 2015; 10(4): 462-3.
[http://dx.doi.org/10.1016/j.jds.2015.03.005]

[208] Yamada Y, Nakamura S, Ito K, *et al.* A feasibility of useful cell-based therapy by bone regeneration with deciduous tooth stem cells, dental pulp stem cells, or bone-marrow-derived mesenchymal stem cells for clinical study using tissue engineering technology. Tissue Eng Part A 2010; 16(6): 1891-900.
[http://dx.doi.org/10.1089/ten.tea.2009.0732] [PMID: 20067397]

[209] Mitani A, Takasu H, Horibe T, *et al.* Five-year clinical results for treatment of intrabony defects with EMD, guided tissue regeneration and open-flap debridement: a case series. J Periodontal Res 2015; 50(1): 123-30.
[http://dx.doi.org/10.1111/jre.12188] [PMID: 24815103]

[210] Liu H-C, e LL, Wang DS, *et al.* Reconstruction of alveolar bone defects using bone morphogenetic protein 2 mediated rabbit dental pulp stem cells seeded on nano-hydroxyapatite/collagen/poly(-lactide). Tissue Eng Part A 2011; 17(19-20): 2417-33.
[http://dx.doi.org/10.1089/ten.tea.2010.0620] [PMID: 21563858]

[211] Chung VH, Chen AY, Kwan C-C, Chen PK, Chang SC. Mandibular alveolar bony defect repair using bone morphogenetic protein 2-expressing autologous mesenchymal stem cells. J Craniofac Surg 2011; 22(2): 450-4.
[http://dx.doi.org/10.1097/SCS.0b013e3182077de9] [PMID: 21403565]

[212] Liu H, Li M, Du L, Yang P, Ge S. Local administration of stromal cell-derived factor-1 promotes stem cell recruitment and bone regeneration in a rat periodontal bone defect model. Mater Sci Eng C 2015; 53: 83-94.
[http://dx.doi.org/10.1016/j.msec.2015.04.002] [PMID: 26042694]

CHAPTER 3

Cardiovascular System Tissue Engineering

Kai Zhu[1,2,§], **Margaux Duchamp**[1,5,§], **Julio Aleman**[3,4,§], **Wanting Niu**[6,7], **Abdulmonem Alshihri**[8,9], **Yu Shrike Zhang**[1,*] and **Ming Yan**[10,*]

[1] *Division of Engineering in Medicine, Department of Medicine, Brigham and Women's Hospital, Harvard Medical School, Cambridge, MA, USA*

[2] *Department of Cardiac Surgery, Zhongshan Hospital, Fudan University, Shanghai, China*

[3] *Molecular and Cellular Biosciences, Wake Forest School of Medicine, Winston Salem, NC, USA*

[4] *Wake Forest Institute for Regenerative Medicine, Winston-Salem, NC 27101, USA*

[5] *Department of Bioengineering, Ãcole Polytechnique Fédérale de Lausanne, Lausanne 1015, Switzerland*

[6] *Tissue Engineering Labs, VA Boston Healthcare System, Boston, MA 02130, USA*

[7] *Department of Orthopedics, Brigham and Women's Hospital, Harvard Medical School. Boston 02115, USA*

[8] *Department of Restorative and Biomaterials, Harvard School of Dental Medicine, Boston, MA, 02115, USA*

[9] *School of Dentistry, King Saud University, Riyadh 11545, Saudi Arabia*

[10] *Department of Biomedical Engineering, College of Life Information Science and Instrument Engineering, Hangzhou Dianzi University, Hangzhou 310018, China*

Abstract: Cardiovascular disease is one of the leading causes of death worldwide. Transplantation is the conventional treatment, but it has to be alleviated due to scarcity of donors. The emerging field of cardiac tissue engineering aims to develop innovative strategies for the treatment of cardiovascular diseases, through engineering of biomimetic tissue substitutes. Among the different strategies, scaffold-based approaches hold great promises, such as the use of rationally designed porous scaffolds that will effectively guide the development of seeded cells into the formation of functional cardiac tissues. Properly selected biomaterials used for scaffold fabrication promote the interactions among homogeneous or heterogeneous cell types while they maintain biomechanical properties of the microenvironment. Here we review the state-of-art progress in porous scaffold-based fabrication of cardiac tissues and vascular

* **Corresponding author Yu Shrike Zhang:** Division of Engineering in Medicine, Department of Medicine, Brigham and Women's Hospital, Harvard Medical School, Cambridge, MA, USA; Tel: +1 617 768 8581; Fax: +1 617 768 8221; E-mail: yszhang@research.bwh.harvard.edu

* **Corresponding author Ming Yan:** Department of Biomedical Engineering, College of Life Information Science and Instrument Engineering, Hangzhou Dianzi University, Hangzhou 310018, China; Tel: +86 571 8687 8667; Fax: +86 571 8687 8667; E-mail: yanming@hdu.edu.cn

§ These authors contributed equally as the primary author

Mehdi Razavi (Ed.)
All rights reserved-© 2017 Bentham Science Publishers

grafts. While extensive achievements have been consolidated in the past decade, with further advancements we envision the applications of cardiac tissue engineering in the areas of drug screening, disease modeling, and *in vivo* regenerative therapy.

Keywords: Biochemical and biomechanical signals, Cardiac tissue engineering, Cardiomyocytes, Endothelial cells, Extracellular matrix, myocardium, Porous scaffold, Smooth muscle cells, Stem cells, Tissue engineering vascular grafts.

INTRODUCTION

Cardiac tissue engineering (CTE) and tissue-engineered vascular grafts (TEVGs) have undergone exciting progress in the last decade [1]. Thanks to the tremendous advances in cell biology, materials science, and engineering technologies, CTE and TEVGs are finding widespread applications in promoting structural and functional restoration of injured myocardium, treatment of cardiovascular system diseases (CVDs), or drug discovery. Although tremendous progress has been achieved over the past decade, accurately mimicking the biology and physiology of the native cardiac tissues remains the main challenge in this field. In native myocardium, extracellular matrix (ECM) modulates cell behaviors and coordinates myofibril assemblies, thereby influencing the proper organization of the whole tissue architecture and contractile function [2, 3]. Recapitulation of the cell-incorporated ECM network using porous scaffolds has thus become a crucial design consideration for engineered cardiac tissues [4]. Porous scaffolds are usually characterized by microscale pores of 50-200 μm and large matrix porosity, which can well support the cell infiltration, growth, and cell-cell interactions while enabling vascularization and diffusion of nutrients or metabolic waste products [5 - 7]. Moreover, the porous scaffolds can be fabricated with sufficient mechanical strength to support the initial tissue architecture and subsequently degrade as needed at a desired rate that matches that of new cardiac tissue formation [7, 8]. In combination with the induced pluripotent stem cell (iPSC) technology, the porous scaffold-based CTE holds great potential in regenerative and personalized medicine.

In this chapter, we illustrate the state-of-the-art progress on engineering porous scaffolds for CTE and TEVGs. The cardiovascular tissue-engineered triad consists of scaffolds, cells, and biomechanical considerations that define the grafts and their successful clinical translation. In particular, we will discuss a variety of different biomaterials, including decellularized matrices, natural polymers, and synthetic polymers, used for fabrication of porous cardiac and vascular scaffolds. We will next cover technologies for construction of scaffolds for generation of cardiac tissues to achieve biomimetic vascularization, anisotropy, and electrical conduction, followed by prospective components for TEVGs and their approaches. We finally conclude with *in vitro* and *in vivo* applications of these engineered cardiovascular tissues and future perspectives.

BIOMATERIALS FOR POROUS CARDIAC SCAFFOLDS FABRICATION

Decellularized ECM-Based Porous Scaffolds

The decellularized ECM (dECM) may be the most structurally and compositionally relevant biomaterial for (cardiac) tissue engineering [9, 10]. It has already been successfully applied to fabrication of porous scaffolds in both preclinical animal studies and human clinical applications [10]. For CTE, Taylor and co-workers first reported the generation of decellularized scaffolds through coronary perfusion in intact rat hearts [11]. In their procedure, the aorta was cannulated into the heart chamber (right ventricle) for retrograde perfusion with ionic detergents, thereby obtaining a dECM porous scaffold with an intact and acellular cardiac geometry. The scaffold was then re-seeded with cardiomyocytes, thereby achieving immature heart pump function *in vitro* (Fig. **1A**). Yang and co-workers repopulated decellularized mouse hearts with human induced pluripotent stem cell (iPSC)-derived cardiovascular progenitor cells. Together with the perfusion of exogenous growth factors, the seeded cells differentiated into cardiomyocytes, smooth muscle cells (SMCs), and endothelial cells (ECs), which facilitated the reconstruction of cardiac muscles and promoted re-endothelialization of the heart constructs (Fig. **1B**) [12]. Even though these decellularized scaffolds can preserve the native structure and biochemical components, their use as heart transplants remains far from feasible and is clinically unpractical due to the poor *in vivo* performances [13]. A step further, in a recent work the Ott Group partially recellularized human whole-heart scaffolds with human iPSC-derived cardiomyocytes with the aim of showing that functional myocardial tissue of human scale could be built based on dECM. They demonstrated that the seeded constructs developed force-generating human myocardial tissues with *in vivo*-like electrophysiology, left ventricular pressure development, and even metabolic function [10]. Dahl [14] and Schaner [15] have individually decellularized arterial and vein sections to use as scaffolds for TEVGs. The dECM-based TEVGs could be an effective strategy to provide graft source, without the usage of allogeneic or xenogeneic vessels.

On the other hand, dECM may also be used as biomaterials for generation of small-sized cardiac patches. These porous scaffolds can be fabricated using solubilized dECM as a 3D nanofibrous hydrogel at physiological pH and temperature. However, dECM hydrogels have a limited range of mechanical properties and rapid degradation kinetics, which make them less suitable for cell encapsulation and *in vivo* engraftment [16]. To this end, Black and co-workers used solubilized dECM from two developmental stages (neonatal, adult) combined with fibrin hydrogels that were crosslinked with transglutaminase [17]. The hybrid dECM/hydrogel-based scaffolds possessed a tunable composition and specific elastic moduli to mimic properties of the developing and mature myocardium. In addition, researchers have also devised various methods for scaffold fabrication using dECM. Cho co-workers adopted grinded dECM of heart as the bioink, for direct three-dimensional (3D) bioprinting of porous scaffolds that mimicked the *in vivo* structure and compositions of the myocardium, increasing cardiac cell engraftment, survival, and long-term functionality (Fig. **2**) [18]. Atala and co-workers reported a combination of poly(ethylene glycol) (PEG)-based crosslinker with dECM for bioprinting. Using this approach, the dECM bioinks could be extruded and reach native mechanical strength once bioprinted and crosslinked [19]. Alternatively, Cho and colleagues reported vitamin B2 as an alternative biocompatible photocrosslinking agent in their cardiac tissue-derived dECM bioink, which allowed convenient control over the mechanics of the bioprinted dECM structures [20]. These examples demonstrated that dECM bioinks can be suitable for CTE in resembling the native biochemistry with tunable biomechanics for direct 3D bioprinting of tissue constructs at low-t--no associated toxicity.

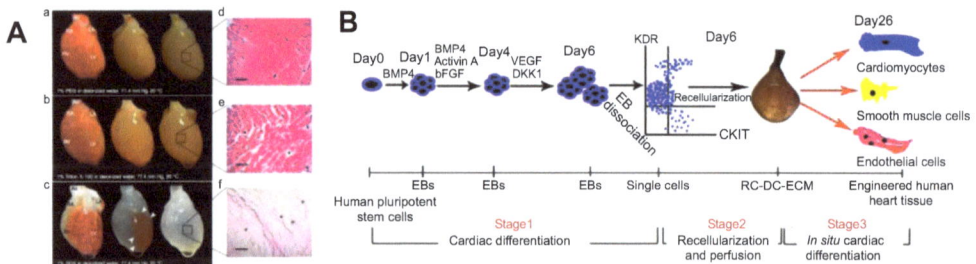

Fig. (1). Repopulation of decellularized heart using human iPSC. (A) Decellularization of whole cadaveric rat heart through retrograde perfusion sequentially with PEG, triton X-100, and sodium dodecyl sulfate that gradually washed out the cellular components. Hematoxylin and eosin (H&E) staining results on thin sections of the left ventricle for each detergent step are shown on the right. Adapted by permission from Nature Publishing Group [11] copyright 2008. (B) Scheme showing the recellularization of mouse hearts with human iPSC-derived multipotent cardiovascular progenitor cells, as well as the *in situ* differentiation over time of these progenitor cells into cardiac lineages. Adapted by permission from Nature Publishing Group [12] copyright 2013.

Polymers-Based Porous Scaffolds

Fabrication of porous scaffolds for tissue engineering applications has ventured into the fields of natural and synthetic polymers as their primary resource for building blocks [21]. Natural polymers suitable for scaffold fabrication such as collagens, gelatin, fibrin, alginate, chitosan, and hyaluronic acid, feature high biocompatibility with intrinsic cell-binding domains and thus enhance cell adhesion and proliferation [22, 23]. They have been processed into various forms of porous structures that are suitable for use in engineering cardiac tissues. However, application of natural polymers have been restrained due to their poor mechanical properties, fast degradation kinetics, and significant variability among different batches of fabrication.

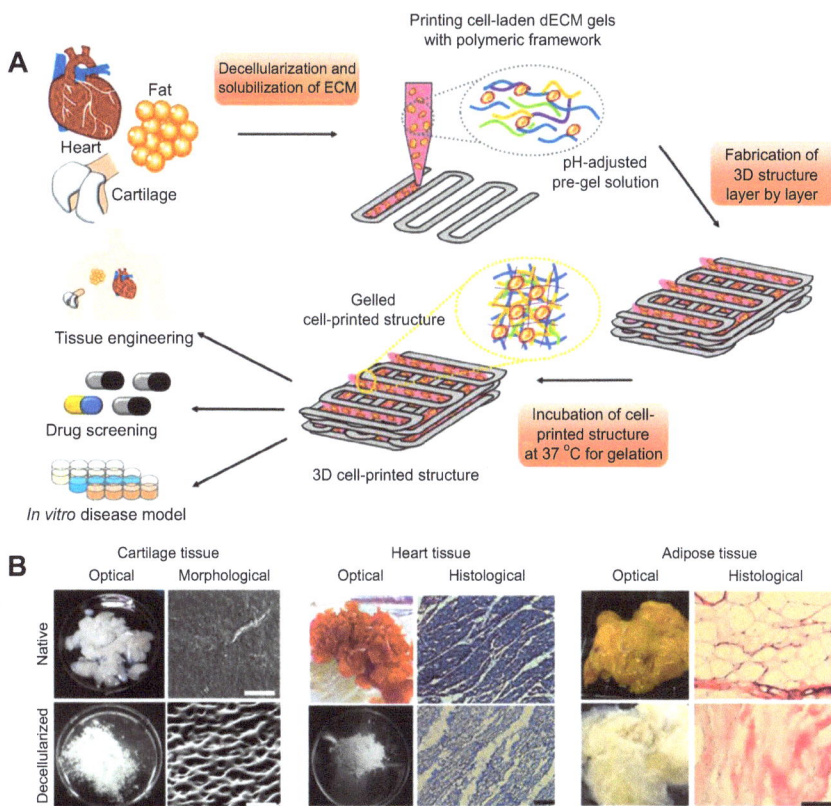

Fig. (2). (A) Schematic detailing the tissue bioprinting process using dECM bioink. The decellularization process combines physical, chemical, and enzymatic processes; the ECM is then solubilized in acidic condition, and adjusted to physiological pH, where living stem cells are embedded inside to achieve 3D printing. (B) Both photographic and microscopic images of native and decellularized cartilage tissue (scale bar, 50 mm), heart tissue (scale bar, 100 mm), and adipose tissue (scale bar, 100 mm). Adapted by permission from Nature Publishing Group [18] copyright 2014.

Synthetic polymers play critical roles as they show high workability, good mechanical properties, and degradation profiles that can all be conveniently and precisely tuned [24]. Moreover, they are easily handled and fabricated into different shapes. Several synthetic polymers and their copolymers, such as polyglycolic acid (PGA), polylactic acid (PLA), and their co-polymers, are approved by U.S. Food and Drug Administration (FDA) for certain clinical applications [25 - 28]. However, their hydrophobicity and the acidic degradation products have constituted serious drawbacks. PEG as a hydrophilic polymer has been used extensively in CTE [29, 30]. Polycaprolactone (PCL) is another versatile biomaterial that is suitable for the design of porous scaffolds. It slowly degrades through hydrolysis within the body and the resultant fragments could be eliminated by the immune system [31].

Furthermore, biomimetic modification of both natural and synthetic polymers can elicit specific cellular functions mediated through biomolecular recognition [32]. To achieve this goal, surface or bulk modifications with bioactive molecules may be incorporated. For example, the arginine-glycine-aspartic acid (RGD) peptide is a biomacromolecule that can promote cell adhesion, prevent cell apoptosis, and accelerate tissue regeneration [33]. The immobilization of RGD peptide onto porous scaffolds would therefore contribute to an enhanced expression of cardiac markers, such as sarcomeric α-actinin, N-cadherin, and connexin-43 [34, 35].

APPROACHES FOR FABRICATION OF POROUS CARDIAC SCAFFOLDS

Several strategies have been investigated for the fabrication of porous scaffolds for CTE. The most common techniques include particulate leaching, foaming process, microsphere sintering, freeze-drying, electrospinning, and soft lithography, among others [36 - 41]. These approaches allow for the fabrication of porous scaffolds from both synthetic polymers and natural materials across multiple length scales [42 - 46]. Nevertheless, they generally suffer from limitations related with structural complexity and reproducibility among different batches of fabrication. While these can be achieved with soft lithography, the procedures are typically complex and lack automation required for large-scale fabrication.

More recently, 3D bioprinting has been introduced into the area of tissue engineering for more flexible and accurate deposition of biomaterials and cells in a precisely defined and automated manner [47]. In general, these technologies can be classified into "top-down" and "bottom-up" approaches [48]. On the one hand for the "top-down" approach, cells are seeded onto a pre-fabricated porous scaffold to induce subsequent creation of their own ECM. Thanks to this strategy, organs of lower levels of complexity, with a few millimeters in size and poor vascularization, have been successfully engineered [49]. However, in order to fabricate organs with higher complexity such as the heart, the main challenge lies in the ability to achieve structural hierarchy suitable to maintain the functionality of multiple cell populations. This inability is mainly due to the limitation of recreating the intricate microstructural features of tissues through the "top-down" strategy [50]. On the other hand, the "bottom-up" strategy may hold potential to address these challenges. It can generate biomimetic tissues *via* the application of microscale building blocks with specific microarchitectures and the assembly of these functional subunits into a larger tissue construct from the bottom up [51]. The disadvantage of the bottom-up approach lies in its slightly complicated fabrication procedure compared to the top-down method. Despite their respective limitations, both technologies have significantly contributed to the engineering of functional cardiac tissues [52 - 59].

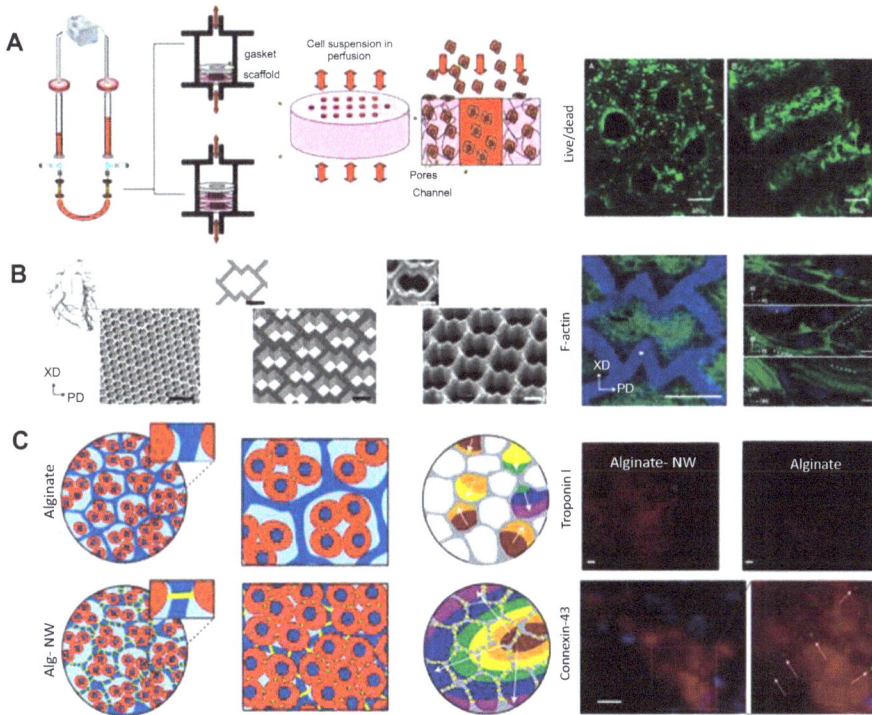

Fig. (3). (A) Designs of perfusion seeding loop, and perfusion cartridge setup both in the single and stacked scaffold configurations. The last scheme shows how the alternation of the cell suspension flow leads to the loading of the cells within the scaffold. Microscopic images on the left show cardiac cell viability in the seeded scaffold (on the left top-down view of the channels while on the right cross-sectional view). Adapted by permission from Wiley-VCH [52] copyright 2010. (B) Schematics and scanning electron microscopy images showing an anisotropic scaffold (left) as well as low- and high-magnification confocal images of graft pores filled with neonatal rat heart cells (f-actin in green and counter nuclear staining in blue) on the right. Adapted by permission from Nature Publishing Group [58] copyright 2008. (C) 3D nanowire cardiac tissue showing the synchronous beating of the cardiac cells when placed in an alginate and nanowire hydrogel. In the top right images immunostaining for troponin I in the cell-seeded scaffolds shows a stronger staining in the alginate-nanowire hydrogel. In the bottom left images the presence of connexin-43 (green) indicates gap junctions were found between the cardiomyocytes. Adapted by permission from Nature Publishing Group [59] copyright 2011.

Vascularization

For multiple applications, the development of a proper vascular network that supplies oxygen and nutrients for tissue formation within a scaffold is highly desired to achieve clinical relevancy. Ischemia prevention is particularly important for the survival and function of engineered cardiac tissues since cardiomyocytes are very sensitive to oxygen concentrations and will only function within a narrow range of oxygen [49]. Interconnected patterns can be formed first inside a scaffold and subsequently seeded with endothelial cells to form the vasculature. For instance, Vunjak-Novakovic and co-workers developed a class of elastomeric poly(glycerol-*co*-sebacate) (PGS) scaffolds containing parallel channels using computerized carbon dioxide laser piercing [52]. The cell-seeded scaffolds were then stacked one-by-one in a perfusion cartridge to achieve uniform spatial distribution of rat cardiomyocytes in the assembled scaffolds. Afterwards, endothelial cells were lined inside the channels *via* transverse perfusion seeding for vascularization (Fig. **3A**). Cohen and co-workers used a CO_2 laser engraving system to form 200-μm-diameter channels through the RGD-alginate scaffolds, with a channel-to-channel distance of 400 μm [60]. After the pre-seeding with endothelial cells and subsequent seeding of cardiac cells, the vessel-like networks in the construct was formed by the endothelial cells remaining in and around the channels due to their interactions with the matrix-bound RGD peptides, while the cardiomyocytes were located in the regions between the channels.

Anisotropy

In the ventricular myocardium, myocardial fibers are interwoven within the collagen sheaths, *i.e.*, perimysial fibers, which are undulated to form a honeycomb-like network [53]. These features yield directionally dependent electrical and mechanical properties collectively termed cardiac anisotropy. The bulk-designed porous scaffolds described so far could not support the development of a substantial anisotropic cardiac tissue patch since their interior structure is isotropic in nature. With the aim of achieving anisotropic property, Freed and co-workers created an accordion-like honeycomb microstructure with PGS, yielding elastomeric scaffolds with controllable stiffness and anisotropy (Fig. **3B**) [58]. The authors reported that the accordion-like honeycomb scaffolds exhibiting distinct orthogonal material directions could potentially match the anisotropic in-plane physiologically mechanical response of native myocardium and reduce in-plane resistance to contraction. Moreover, they provided an inherent structural capacity to guide cardiomyocytes orientation in the absence of external stimuli. Ratner *et al.* reported a microtemplating method to shape poly(2-hydroxyethyl methacrylate-co-methacrylic acid) (pHEMMA) hydrogel into a scaffold [56]. The construct contains parallel channels to organize cardiomyocytes into antistrophic bundles and micrometer-sized, spherical, interconnected pores to facilitate the formation of vessel-like network. These engineered scaffolds closely matched the mechanical and contractile properties characteristic of the native adult myocardium.

Cardiac Maturation and Conduction

Biomaterial-based matrices are poor in electrical conductivity, which limits cell-cell interactions and thus will delay electrical signal propagation between cardiomyocytes. Therefore, the engineered cardiac tissues usually possess low contractile strengths compared with native myocardium. In an effort to overcome this disadvantage, Kohane and co-workers reported a method to incorporate gold nanowires with an average length of approximately 1 μm into alginate scaffolds, in order to bridge the electrically resistant polymer chains (Fig. **3C**) [59]. The gold-alginate composite scaffolds showed improved mechanical properties and electrical conductivity. Cardiac cells (cardiomyocytes and cardiac fibroblasts) populated on the conductive scaffolds indicated thicker growth and better alignment than those seeded on pristine alginate scaffolds. When electrically stimulated, cardiomyocytes in these tissues contracted synchronously, as evaluated by calcium imaging showing calcium transient signals at various sites in the conductive scaffolds, compared to calcium transients only at the stimulation site in pristine scaffolds. Furthermore, higher levels of proteins involved in cardiac muscle contraction and electrical coupling (troponin I and connexin-43) were detected in the composite matrices. Besides gold nanostructures (*e.g.* nanoparticles, nanorods, and nanowires) [59, 61, 62], other conductive materials that have been integrated with polymer scaffolds include carbon nanotubes and (reduced) graphene oxide [63, 64].

APPLICATIONS OF POROUS SCAFFOLDS IN CTE

In Vitro Modeling

Porous scaffolds with cardiac tissue constructs can provide *in vitro* platforms for bioactuator (Fig. **4A**) [63], drug discovery, biological assessment, and disease modeling (Fig. **4B**) [65]. Drug-induced cardiotoxicity is a great concern for drug discovery in pharmaceutical industries. Drug tests on animals are expensive, time-consuming, associated with ethical concerns, and differ from the physiological responses in humans [66]. In the last decade, the availability to obtain human cardiomyocytes either derived from human embryonic stem cells (ESCs) or iPSCs has enabled personalized drug screening [67]. Compared to two-dimensional (2D) models, tissue-engineering techniques deliver a 3D biomimetic cell-laden scaffold that behaves closer to *in vivo* physiological conditions. These miniaturized *in vitro* heart models can be used to probe the systemic side effects of pharmaceutical compounds on the human cardiac tissues and therefore better predict drug responses in the human body [68]. Another application of the *in vitro* cardiac models is biological assessment and modeling of cardiac diseases, through the use of cardiomyocytes derived from diseased heart tissues or iPSCs [69]. These *in vitro* mature cardiac scaffolds may be used as platforms to more accurately investigate physiology of normal cardiac tissues or heart pathologies.

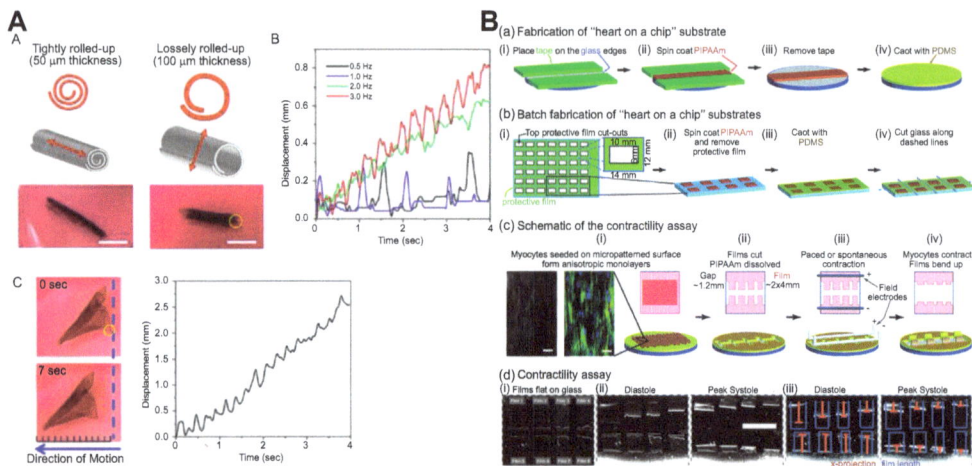

Fig. (4). (A) Engineered 3D biohybrid actuators from free-standing cardiac patches: (a) Schematics of carbon nanotube-hydrogel-based cardiac scaffold for construction of bioactuators (tightly and loosely rolled-up forms) and their corresponding beating directions (red arrow). (b) Displacement of the bioactuator over time under various frequencies. (c) Optical images (at 0 and 7 s) of the spontaneous linear traveling of a triangular swimmer with the respective displacement *versus* time plot. Adapted by permission from American Chemical Society [63] copyright 2013. (B) Assembly of heart-on-a-chip platform with the development of microscale tissue constructs for *in vitro* test with (a) first the fabrication of the substrate and (b) with afterwards the batch fabrication. The contractility assay is described with (c) a schematic and (d) the obtained results under various conditions. Adapted with permission from Royal Society of Chemistry [65] copyright 2011.

In Vivo Therapy

CTE also aims at providing living, force-producing cardiac tissues that can be transplanted to injured or malformed hearts to restore cardiac functions (Fig. **5**) [70]. A key parameter of scaffolds with possible *in vivo* applications is the degradation rate, since scaffolds should not degrade shortly after engraftment, as they are necessary to supply a platform for initial cell growth. On the other hand, the scaffolds should degrade at a suitable rate to be gradually replaced by the newly generated cardiac ECM and functional myocardium. Immune response to xenogeneic or allogeneic materials that might inherent to the construct should be examined before moving into scaffold engraftment *in vivo*. Efforts to reduce graft rejection including anti-inflammatory implementation; biologically derived polymers and synthetic anti-inflammatory drugs have been incorporated into porous scaffolds when engineering cardiac tissues [71]. Currently, a few studies have been performed in large animal myocardium infarction models [72, 73]. In addition, many promising results have been obtained in large animal-based preclinical studies, and few clinical trials have tested scaffolds with or without cells (ClinicalTrials.gov Identifier: NCT00981006, NCT02139189, and NCT02057900). Preclinical and clinical trials that overcome current challenges associated with existing CTE techniques will enable a better translation of such constructs to future clinical applications.

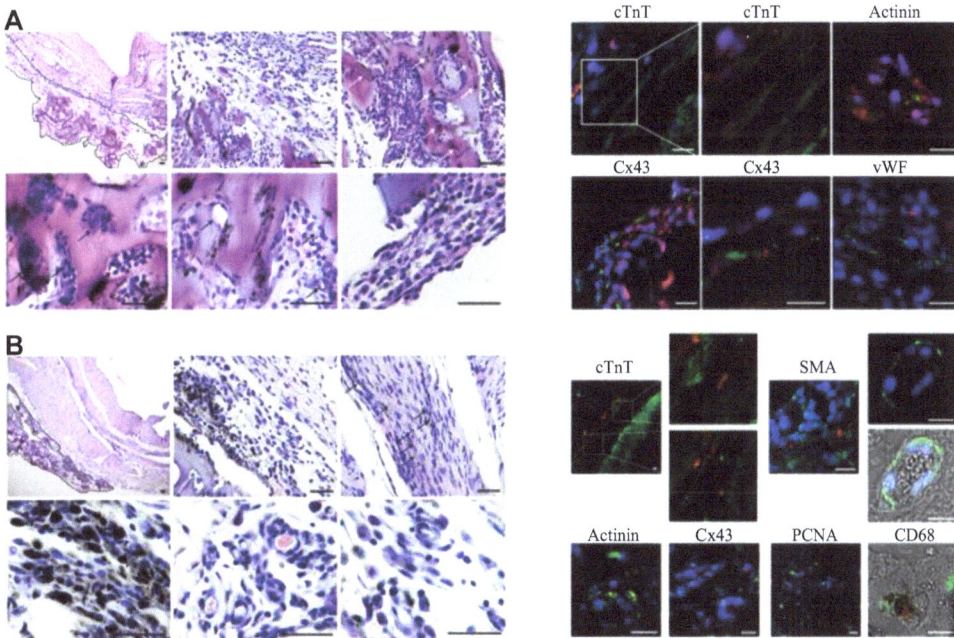

Fig. (5). Morphology of the cardiac scaffolds after engraftment. (A) Images obtained 1 week after the engraftment of H&E staining (on the left) showing the attachment of the hydrogel scaffold to the infarct surface (black dotted line) as well as immunostainings (on the right) indicating a differentiated phenotype. (B) Images after 4 weeks of engraftment, where the H&E staining (on the left) highlights the formation of blood vessels, and migration of cardiomyocytes into the scar area. Adapted by permission from Nature Publishing Group [70] copyright 2014.

TISSUE-ENGINEERED VASCULAR GRAFTS

TEVGs require a biocompatible scaffold, a reliable source of appropriate cells, and proper coordination of physicochemical parameters for proper tissue function.

Blood Vessel Physiology

All vessels *in vivo* are coated with a layer of confluent ECs termed the endothelium, which conforms the tunica intima layer. The vascular ECs play crucial roles in regulation of vascular homeostasis. For example, they synthesize and secret a variety of vasoactive agents to regulate vasomotor tone, angiogenesis, thrombosis, and atherogenesis [74]. Therefore, a minimum requirement for TEVGs, should be an endothelial layer covering the inner side of tubular vascular scaffolds. There are two main strategies utilized to promote endothelialization: the *in vitro* endothelialization and methods for *in vivo* endothelialization. *In vitro* efforts seed ECs to the scaffolds that are then implanted into the injury site, while the *in vivo* endothelialization requires no initial coating of ECs; instead, the graft is pretreated with either ECM components such as gelatin, fibrin, collagen, or fibronectin that actively recruit endothelial cells to restore the integrity of the endothelium. Factors such as vascular endothelial growth factor (VEGF), basic fibroblast growth factor (bFGF), and the hepatocyte growth factor (HGF) promote migration as well. However, the limited capacity in the regeneration of ECs becomes an obstacle to the successful endothelialization of TEVGs. There has been limited EC ingrowth in currently available grafts. It is urgent to develop better strategies to address these unresolved challenges [75].

Outside the intima is the basement membrane, the main components of which include type IV collagen, laminin, proteoglycan, and fibronectin. The media is composed of collagens types I and III, as well as embedded SMCs [76]. SMCs receive stimulating signals in the form of cytokines and then contract or dilate in a coordinated fashion, leading to peristaltic movement of the vessel [77]. As a result, TEVGs should also contain circumferentially arranged contractile SMCs to recapitulate vasoactivity. In normal blood vessels, SMCs proliferate at a low rate and are suitable for contraction. For TEVGs scaffolds, seeded SMCs must have plasticity capable of switching between a contractile phenotype and a temporary phenotype with higher proliferation rate and ECM secretion capability. However, excessive proliferation in the implanted graft may result in vessel wall thickening, vascular stenosis, and the subsequent failure of the grafts [78].

Porous Scaffold Design

The design of porous scaffolds for TEVGs is largely similar to that for CTE, including the use of both natural and synthetic polymers, as well as the different approaches adopted for fabrication of the porous scaffolds (see Sections 2.2 and 3). As with CTE, decellularization methods have also been developed for engineering vascular grafts [79 - 81]. For example, DiMuzio and co-workers decellularized intact human greater saphenous veins leading to >94% removal of cells but retaining of the dECM molecules such as collagens [15]. The decellularized veins when implanted into recipient animals showed satisfactory functionality. Another example falls on to the strategy devised by Langer and co-workers back in the late 1990s, who fabricated vascular replacements by growing vascular SMCs within processed porous tubular polymeric scaffolds *in vitro* followed by decellularizing them [26]. The company Humacyte was subsequently founded to continue the development of such type of TEVG, which has recently entered clinical phase trials in human patients (ClinicalTrials.gov Identifier: NCT02644941).

Cells Sources

While mature cell types can directly assume vascular functions, their sources are limited and scalable production remains challenging. To this end, stem cells have been widely used in TEVGs. For example, due to their ability to differentiate into multiple tissue lineages, mesenchymal stem cells (MSCs) have been well studied in the vascular tissue engineering [82, 83]. The typical morphology of MSCs is fibroblastic and spindle-shaped. The main source of MSCs is isolation by aspiration of the iliac crest or from other non-marrow-derived sources [84]. However, the differentiation of MSCs towards SMCs and ECs is affected by many factors, including growth factors, mechanical stimuli, and scaffolding technology. Upon differentiation, the cells incorporated into an engineered graft should be further tested *in vitro* or implanted *in vivo* [85]. In addition, stem cells such as embryonic stem cells (ES) also represent a possibility as progenitor cells for vascular engineering, as differentiation of human ES cells into SMCs and EC lines have been successfully achieved [86, 87].

Signaling Factors

The targeted and controlled stimulation of select cells through biochemical and mechanical signals is another important aspect. It is critical to drive the formation of new tissue for the therapeutic replacement of the impaired blood vessels [88].

Shear Stress

Because of its prominent role in pathogenesis of cardiovascular diseases such as atherosclerosis, shear stress has gained increasing attention in recent years. Both *in vitro* and *in vivo* investigations have demonstrated that vascular ECs alter their production of vasoactive substances and the expression of adhesion molecules under shear stress, which may further affect vascular remodeling [89, 90]. Hence, to provide a shear-stable EC layer, strategies such as coating vascular grafts with the peptides that promote ECs adhesion [91] or immobilization of growth factors to TEVGs have been used to promote endothelialization of vascular grafts [92, 93]. Their success showed great promise to increase the patency of synthetic small diameter vascular grafts.

Mechanical Stretch

Due to the pulsatile blood flow, blood vessels are subjected to mechanical forces in a form of radial distention. Several groups have reported that stretching cells seeded on the scaffolds can induce changes in cell phenotype, cell differentiation and the production of ECM proteins [94, 95]. Applying cyclic strain to a vascular graft causes vascular SMCs on it to switch to a contractile phenotype, which is critical for contractile functions of SMCs [96]. Cyclical stretch could also cause changes in ECM organization and the mechanical properties of TEVGs [97, 98]. The TEVG should thus receive proper biomechanical signals for successful functional generation and engraftment [88].

Biochemical Factors

The local cellular microenvironment, composed of soluble growth factors, neighboring cells, and surrounding ECMs, dynamically regulate cell functions. A strategy for vascular regeneration, therefore, is providing suitable biochemical factors to mimic the microenvironment of vascular cells. However, the biochemical component of the culture environment is considerably more complicated to optimize than is the mechanical component.

In general, there are two classes of biochemical factors in TEVGs: *i*) biochemicals used for ECM production and stabilization, and *ii*) cell signaling molecules such as growth factors that are believed to be therapeutically effective for neotissue development and the healing process. For example, supplementing growth factors in the culture media could drive desired responses from the target cells before implant. Another strategy is the localized delivery of growth factors. Researchers developed a variety of methods to covalently immobilize growth factors into a biomaterial matrix [99]. The dual release of VEGF and platelet-derived growth factor (PDGF) from an electrospun multilayered small-diameter vascular scaffold significantly promoted endothelialization and inhibited the hyperproliferation of SMCs seeded on top, potentially benefiting blood vessel reconstruction.

CONCLUSIONS AND PERSPECTIVES

Significant advances in our structural and mechanical knowledge of the heart, as well as improvements in the materials production methods and engineering technologies used to fabricate porous scaffolds have been made during the past decade for CTE-based heart regeneration. Among the different parameters, structural properties of the scaffold, surface properties, mechanical properties, and electrical properties represent critical elements for successfully engineering cardiac tissues. Moreover, the integration of these parameters with the co-culture of multiple cell types constituting the heart, are among the minimum requirements to achieve the most successful CTE construct to date.

Assessment of cell infiltration, scaffold degradation, degree of vascularization and remodeling can be conventionally accomplished with histological analysis following *in vivo* transplantation. However, this type of end-point analysis often requires termination of the scaffold/tissue constructs. Implementation of real-time, non-invasive imaging and functional analysis methodologies capable of assessing engineered tissues *in situ* will enable the assessment of true time-dependent nature of the remodeling. Clinical diagnostic tools such as magnetic resonance imaging techniques (MRI), computerized tomography (CT), fluorescence imaging techniques, and echo/Doppler techniques may be useful to track scaffold/tissue growth and integration [100 - 103]. Additionally, scaffolds manufactured with embedded dopants such as magnetic or conductive nanoparticles may potentially aid external, non-destructive monitoring of scaffold/tissue interactions.

Earlier studies focused on animal-derived cardiomyocytes due to their easy accessibility, the paradigm is now starting to shift toward the use of cardiomyocytes of human origin, particularly those derived from iPSCs [72]. Indeed, the fact that human iPSCs are derived in a patient-matched manner has made them an exquisite cell source for precise heart models that can be used for personalized drug screening, understanding patient-specific fundamentals of diseases, as well as to minimize immune and fibrotic responses in *in vivo* graft transplantation.

Engineered porous scaffolds have been used in patients for skin, cartilage, bladder, and vascular grafts [41]. In comparison, the translation in the field of cardiac scaffolds has progressed at a slower pace. One of the major reasons lies in the limitation of delivery approach where scaffolds have been usually sutured or attached onto the epicardial surface of the heart muscle during an open-heart surgery. In the future, we envision a further minimally invasive delivery approach, such as video-assisted thoracic surgery and robotic surgery among others. *In situ* delivery of enhanced scaffolds with high myocardial integrity will be a regular clinical procedure. Regeneration through the path of cardiac engineering has opened paths towards achieving successful repair of myocardial damages, as a novel way to build a heart.

ABBREVIATIONS

CTE	cardiac tissue engineering
CVD	cardiovascular system disease
dECM	decellularized extracellular matrix
ECM	extracellular matrix

ESC	embryonic stem cell
GF	growth factor
hESC	human embryonic stem cell
iPSC	induced pluripotent stem cell
MSC	mesenchymal stem cell
PCL	polycaprolactone
PDGF	platelet-derived growth factor
pHEMMA	poly(2-hydroxyethyl methacrylate-*co*-methacrylic acid)
PGA	polyglycolic acid
PGS	poly(glycerol-*co*-sebacate)
PLA	polylactic acid
RGD	Arginine-glycine-aspartic acid
SMC	smooth muscle cell
TEVG	tissue engineering vascular graft
VEGF	vascular endothelial growth factor

CONFLICT OF INTEREST

The authors declare no conflict of interest, financial or otherwise.

ACKNOWLEDGEMENT

K.Z. acknowledges National Natural Science Foundation of China (81301312) and "Chen Guang" project supported by Shanghai Municipal Education Commission and Shanghai Education Development Foundation (14CG06). Y.S.Z. acknowledges the National Cancer Institute of the National Institutes of Health Pathway to Independence Award (K99CA201603). M.Y. acknowledges Zhejiang Provincial Natural Science Foundation (No. LY15H180012) and National Natural Science Foundation of China (No. 30900301).

REFERENCES

[1]   Hirt MN, Hansen A, Eschenhagen T. Cardiac tissue engineering: state of the art. Circ Res 2014; 114(2): 354-67.
[http://dx.doi.org/10.1161/CIRCRESAHA.114.300522] [PMID: 24436431]

[2]   Dobaczewski M, Gonzalez-Quesada C, Frangogiannis NG. The extracellular matrix as a modulator of the inflammatory and reparative response following myocardial infarction. J Mol Cell Cardiol 2010; 48(3): 504-11.
[http://dx.doi.org/10.1016/j.yjmcc.2009.07.015] [PMID: 19631653]

[3]   Chen J-H, Simmons CA. Cell-matrix interactions in the pathobiology of calcific aortic valve disease: critical roles for matricellular, matricrine, and matrix mechanics cues. Circ Res 2011; 108(12): 1510-24.

[http://dx.doi.org/10.1161/CIRCRESAHA.110.234237] [PMID: 21659654]

[4]   Parker KK, Ingber DE. Extracellular matrix, mechanotransduction and structural hierarchies in heart tissue engineering. Philos Trans R Soc Lond B Biol Sci 2007; 362(1484): 1267-79.
        [http://dx.doi.org/10.1098/rstb.2007.2114] [PMID: 17588874]

[5]   Liu Q, Tian S, Zhao C, *et al.* Porous nanofibrous poly(L-lactic acid) scaffolds supporting cardiovascular progenitor cells for cardiac tissue engineering. Acta Biomater 2015; 26: 105-14.
        [http://dx.doi.org/10.1016/j.actbio.2015.08.017] [PMID: 26283164]

[6]   Hollister SJ. Porous scaffold design for tissue engineering. Nat Mater 2005; 4(7): 518-24.
        [http://dx.doi.org/10.1038/nmat1421] [PMID: 16003400]

[7]   Boffito M, Sartori S, Ciardelli G. Polymeric scaffolds for cardiac tissue engineering: requirements and fabrication technologies. Polym Int 2014; 63(1): 2-11.
        [http://dx.doi.org/10.1002/pi.4608]

[8]   Loh QL, Choong C. Three-dimensional scaffolds for tissue engineering applications: role of porosity and pore size. Tissue Eng Part B Rev 2013; 19(6): 485-502.
        [http://dx.doi.org/10.1089/ten.teb.2012.0437] [PMID: 23672709]

[9]   Woo JS, Fishbein MC, Reemtsen B. Histologic examination of decellularized porcine intestinal submucosa extracellular matrix (CorMatrix) in pediatric congenital heart surgery. Cardiovasc Pathol 2016; 25(1): 12-7.
        [http://dx.doi.org/10.1016/j.carpath.2015.08.007] [PMID: 26453090]

[10]   Guyette JP, Charest JM, Mills RW, *et al.* Bioengineering human myocardium on native extracellular matrix. Circ Res 2016; 118(1): 56-72.
        [http://dx.doi.org/10.1161/CIRCRESAHA.115.306874] [PMID: 26503464]

[11]   Ott HC, Matthiesen TS, Goh S-K, *et al.* Perfusion-decellularized matrix: using nature's platform to engineer a bioartificial heart. Nat Med 2008; 14(2): 213-21.
        [http://dx.doi.org/10.1038/nm1684] [PMID: 18193059]

[12]   Lu T-Y, Lin B, Kim J, *et al.* Repopulation of decellularized mouse heart with human induced pluripotent stem cell-derived cardiovascular progenitor cells. Nat Commun 2013; 4: 2307.
        [http://dx.doi.org/10.1038/ncomms3307] [PMID: 23942048]

[13]   Ott HC, Mathisen DJ. Bioartificial tissues and organs: are we ready to translate? Lancet 2011; 378(9808): 1977-8.
        [http://dx.doi.org/10.1016/S0140-6736(11)61791-1] [PMID: 22119608]

[14]   Dahl SL, Koh J, Prabhakar V, Niklason LE. Decellularized native and engineered arterial scaffolds for transplantation. Cell Transplant 2003; 12(6): 659-66.
        [http://dx.doi.org/10.3727/000000003108747136] [PMID: 14579934]

[15]   Schaner PJ, Martin ND, Tulenko TN, *et al.* Decellularized vein as a potential scaffold for vascular tissue engineering. J Vasc Surg 2004; 40(1): 146-53.
        [http://dx.doi.org/10.1016/j.jvs.2004.03.033] [PMID: 15218475]

[16]   Eitan Y, Sarig U, Dahan N, Machluf M. Acellular cardiac extracellular matrix as a scaffold for tissue engineering: *in vitro* cell support, remodeling, and biocompatibility. Tissue Eng Part C Methods 2010; 16(4): 671-83.
        [http://dx.doi.org/10.1089/ten.tec.2009.0111] [PMID: 19780649]

[17]   Williams C, Budina E, Stoppel WL, *et al.* Cardiac extracellular matrix-fibrin hybrid scaffolds with tunable properties for cardiovascular tissue engineering. Acta Biomater 2015; 14: 84-95.
        [http://dx.doi.org/10.1016/j.actbio.2014.11.035] [PMID: 25463503]

[18]   Pati F, Jang J, Ha D-H, *et al.* Printing three-dimensional tissue analogues with decellularized extracellular matrix bioink. Nat Commun 2014; 5: 3935.
        [http://dx.doi.org/10.1038/ncomms4935] [PMID: 24887553]

[19]   Skardal A, Devarasetty M, Kang HW, *et al.* Bioprinting Cellularized Constructs Using a Tissue-specific Hydrogel Bioink 2016.
       [http://dx.doi.org/10.3791/53606]

[20]   Jang J, Kim TG, Kim BS, Kim S-W, Kwon S-M, Cho D-W. Tailoring mechanical properties of decellularized extracellular matrix bioink by vitamin B2-induced photo-crosslinking. Acta Biomater 2016; 33: 88-95.
       [http://dx.doi.org/10.1016/j.actbio.2016.01.013] [PMID: 26774760]

[21]   Jafari M, Paknejad Z, Rad MR, *et al.* Polymeric scaffolds in tissue engineering: a literature review. J Biomed Mater Res B Appl Biomater 2015.
       [PMID: 26496456]

[22]   Stoppel WL, Ghezzi CE, McNamara SL, Black LD III, Kaplan DL. Clinical applications of naturally derived biopolymer-based scaffolds for regenerative medicine. Ann Biomed Eng 2015; 43(3): 657-80.
       [http://dx.doi.org/10.1007/s10439-014-1206-2] [PMID: 25537688]

[23]   Ma L, Gao C, Mao Z, *et al.* Collagen/chitosan porous scaffolds with improved biostability for skin tissue engineering. Biomaterials 2003; 24(26): 4833-41.
       [http://dx.doi.org/10.1016/S0142-9612(03)00374-0] [PMID: 14530080]

[24]   Malafaya PB, Silva GA, Reis RL. Natural-origin polymers as carriers and scaffolds for biomolecules and cell delivery in tissue engineering applications. Adv Drug Deliv Rev 2007; 59(4-5): 207-33.
       [http://dx.doi.org/10.1016/j.addr.2007.03.012] [PMID: 17482309]

[25]   Nair LS, Laurencin CT. Biodegradable polymers as biomaterials. Prog Polym Sci 2007; 32(8): 762-98.
       [http://dx.doi.org/10.1016/j.progpolymsci.2007.05.017]

[26]   Niklason LE, Gao J, Abbott WM, *et al.* Functional arteries grown *in vitro.* Science 1999; 284(5413): 489-93.
       [http://dx.doi.org/10.1126/science.284.5413.489] [PMID: 10205057]

[27]   Hajiali H, Shahgasempour S, Naimi-Jamal MR, Peirovi H. Electrospun PGA/gelatin nanofibrous scaffolds and their potential application in vascular tissue engineering. Int J Nanomedicine 2011; 6: 2133-41.
       [http://dx.doi.org/10.2147/IJN.S24312] [PMID: 22114477]

[28]   Holder WD Jr, Gruber HE, Roland WD, Moore AL, Culberson CR, Loebsack AB, *et al.* Increased vascularization and heterogeneity of vascular structures occurring in polyglycolide matrices containing aortic endothelial cells implanted in the rat. Tissue Eng 1997; 3(2): 149-60.
       [http://dx.doi.org/10.1089/ten.1997.3.149]

[29]   DeLong SA, Moon JJ, West JL. Covalently immobilized gradients of bFGF on hydrogel scaffolds for directed cell migration. Biomaterials 2005; 26(16): 3227-34.
       [http://dx.doi.org/10.1016/j.biomaterials.2004.09.021] [PMID: 15603817]

[30]   Wang H, Feng Y, Fang Z, Yuan W, Khan M. Co-electrospun blends of PU and PEG as potential biocompatible scaffolds for small-diameter vascular tissue engineering. Mater Sci Eng C 2012; 32(8): 2306-15.
       [http://dx.doi.org/10.1016/j.msec.2012.07.001]

[31]   Pektok E, Nottelet B, Tille J-C, *et al.* Degradation and healing characteristics of small-diameter poly(ε-caprolactone) vascular grafts in the rat systemic arterial circulation. Circulation 2008; 118(24): 2563-70.
       [http://dx.doi.org/10.1161/CIRCULATIONAHA.108.795732] [PMID: 19029464]

[32]   Pasparakis G, Krasnogor N, Cronin L, Davis BG, Alexander C. Controlled polymer synthesis--from biomimicry towards synthetic biology. Chem Soc Rev 2010; 39(1): 286-300.
       [http://dx.doi.org/10.1039/B809333B] [PMID: 20023853]

[33]   Liu S. Radiolabeled cyclic RGD peptides as integrin α(v)β(3)-targeted radiotracers: maximizing binding affinity *via* bivalency. Bioconjug Chem 2009; 20(12): 2199-213.

[http://dx.doi.org/10.1021/bc300167c] [PMID: 19719118]

[34]   Hersel U, Dahmen C, Kessler H. RGD modified polymers: biomaterials for stimulated cell adhesion and beyond. Biomaterials 2003; 24(24): 4385-415.
[http://dx.doi.org/10.1016/S0142-9612(03)00343-0] [PMID: 12922151]

[35]   Shachar M, Tsur-Gang O, Dvir T, Leor J, Cohen S. The effect of immobilized RGD peptide in alginate scaffolds on cardiac tissue engineering. Acta Biomater 2011; 7(1): 152-62.
[http://dx.doi.org/10.1016/j.actbio.2010.07.034] [PMID: 20688198]

[36]   Garg T, Goyal AK. Biomaterial-based scaffolds--current status and future directions. Expert Opin Drug Deliv 2014; 11(5): 767-89.
[http://dx.doi.org/10.1517/17425247.2014.891014] [PMID: 24669779]

[37]   Zhang YS, Choi S-W, Xia Y. Inverse opal scaffolds for applications in regenerative medicine. Soft Matter 2013; 9(41): 9747-54.
[http://dx.doi.org/10.1039/c3sm52063c]

[38]   Zhang YS, Xia Y. Multiple facets for extracellular matrix mimicking in regenerative medicine. Nanomedicine (Lond) 2015; 10(5): 689-92.
[http://dx.doi.org/10.2217/nnm.15.10] [PMID: 25816873]

[39]   Ma PX. Biomimetic materials for tissue engineering. Adv Drug Deliv Rev 2008; 60(2): 184-98.
[http://dx.doi.org/10.1016/j.addr.2007.08.041] [PMID: 18045729]

[40]   Rice JJ, Martino MM, De Laporte L, Tortelli F, Briquez PS, Hubbell JA. Engineering the regenerative microenvironment with biomaterials. Adv Healthc Mater 2013; 2(1): 57-71.
[http://dx.doi.org/10.1002/adhm.201200197] [PMID: 23184739]

[41]   Hollister SJ. Scaffold design and manufacturing: from concept to clinic. Adv Mater 2009; 21(32-33): 3330-42.
[http://dx.doi.org/10.1002/adma.200802977] [PMID: 20882500]

[42]   O'Brien FJ, Harley BA, Yannas IV, Gibson L. Influence of freezing rate on pore structure in freeze-dried collagen-GAG scaffolds. Biomaterials 2004; 25(6): 1077-86.
[http://dx.doi.org/10.1016/S0142-9612(03)00630-6] [PMID: 14615173]

[43]   Hu X, Shen H, Yang F, Bei J, Wang S. Preparation and cell affinity of microtubular orientation-structured PLGA(70/30) blood vessel scaffold. Biomaterials 2008; 29(21): 3128-36.
[http://dx.doi.org/10.1016/j.biomaterials.2008.04.010] [PMID: 18439673]

[44]   Agarwal S, Wendorff JH, Greiner A. Use of electrospinning technique for biomedical applications. Polymer (Guildf) 2008; 49(26): 5603-21.
[http://dx.doi.org/10.1016/j.polymer.2008.09.014]

[45]   Xu C, Inai R, Kotaki M, Ramakrishna S. Electrospun nanofiber fabrication as synthetic extracellular matrix and its potential for vascular tissue engineering. Tissue Eng 2004; 10(7-8): 1160-8.
[http://dx.doi.org/10.1089/ten.2004.10.1160] [PMID: 15363172]

[46]   Boland ED, Matthews JA, Pawlowski KJ, Simpson DG, Wnek GE, Bowlin GL. Electrospinning collagen and elastin: preliminary vascular tissue engineering. Front Biosci 2004; 9(1422): 1422-32.
[http://dx.doi.org/10.2741/1313] [PMID: 14977557]

[47]   O'Brien CM, Holmes B, Faucett S, Zhang LG. Three-dimensional printing of nanomaterial scaffolds for complex tissue regeneration. Tissue Eng Part B Rev 2015; 21(1): 103-14.
[http://dx.doi.org/10.1089/ten.teb.2014.0168] [PMID: 25084122]

[48]   Shapira A, Kim D-H, Dvir T. Advanced micro- and nanofabrication technologies for tissue engineering. Biofabrication 2014; 6(2): 020301.
[http://dx.doi.org/10.1088/1758-5082/6/2/020301] [PMID: 24876336]

[49]   Novosel EC, Kleinhans C, Kluger PJ. Vascularization is the key challenge in tissue engineering. Adv Drug Deliv Rev 2011; 63(4-5): 300-11.

[http://dx.doi.org/10.1016/j.addr.2011.03.004] [PMID: 21396416]

[50] Lu T, Li Y, Chen T. Techniques for fabrication and construction of three-dimensional scaffolds for tissue engineering 2013.
[http://dx.doi.org/10.2147/IJN.S38635]

[51] Elbert DL. Bottom-up tissue engineering. Curr Opin Biotechnol 2011; 22(5): 674-80.
[http://dx.doi.org/10.1016/j.copbio.2011.04.001] [PMID: 21524904]

[52] Maidhof R, Marsano A, Lee EJ, Vunjak-Novakovic G. Perfusion seeding of channeled elastomeric scaffolds with myocytes and endothelial cells for cardiac tissue engineering. Biotechnol Prog 2010; 26(2): 565-72.
[PMID: 20052737]

[53] Macchiarelli G, Ohtani O, Nottola S, Stallone T, Camboni A, Prado I, *et al.* A micro-anatomical model of the distribution of myocardial endomysial collagen 2002.

[54] Sun X, Altalhi W, Nunes SS. Vascularization strategies of engineered tissues and their application in cardiac regeneration. Adv Drug Deliv Rev 2016; 96: 183-94.
[http://dx.doi.org/10.1016/j.addr.2015.06.001] [PMID: 26056716]

[55] Capulli AK, MacQueen LA, Sheehy SP, Parker KK. Fibrous scaffolds for building hearts and heart parts. Adv Drug Deliv Rev 2016; 96: 83-102.
[http://dx.doi.org/10.1016/j.addr.2015.11.020] [PMID: 26656602]

[56] Madden LR, Mortisen DJ, Sussman EM, *et al.* Proangiogenic scaffolds as functional templates for cardiac tissue engineering. Proc Natl Acad Sci USA 2010; 107(34): 15211-6.
[http://dx.doi.org/10.1073/pnas.1006442107] [PMID: 20696917]

[57] Balint R, Cassidy NJ, Cartmell SH. Conductive polymers: towards a smart biomaterial for tissue engineering. Acta Biomater 2014; 10(6): 2341-53.
[http://dx.doi.org/10.1016/j.actbio.2014.02.015] [PMID: 24556448]

[58] Engelmayr GC Jr, Cheng M, Bettinger CJ, Borenstein JT, Langer R, Freed LE. Accordion-like honeycombs for tissue engineering of cardiac anisotropy. Nat Mater 2008; 7(12): 1003-10.
[http://dx.doi.org/10.1038/nmat2316] [PMID: 18978786]

[59] Dvir T, Timko BP, Brigham MD, *et al.* Nanowired three-dimensional cardiac patches. Nat Nanotechnol 2011; 6(11): 720-5.
[http://dx.doi.org/10.1038/nnano.2011.160] [PMID: 21946708]

[60] Zieber L, Or S, Ruvinov E, Cohen S. Microfabrication of channel arrays promotes vessel-like network formation in cardiac cell construct and vascularization *in vivo*. Biofabrication 2014; 6(2): 024102.
[http://dx.doi.org/10.1088/1758-5082/6/2/024102] [PMID: 24464741]

[61] Shevach M, Fleischer S, Shapira A, Dvir T. Gold nanoparticle-decellularized matrix hybrids for cardiac tissue engineering. Nano Lett 2014; 14(10): 5792-6.
[http://dx.doi.org/10.1021/nl502673m] [PMID: 25176294]

[62] Navaei A, Saini H, Christenson W, Sullivan RT, Ros R, Nikkhah M. Gold nanorod-incorporated gelatin-based conductive hydrogels for engineering cardiac tissue constructs. Acta Biomater 2016; 41: 133-46.
[http://dx.doi.org/10.1016/j.actbio.2016.05.027] [PMID: 27212425]

[63] Shin SR, Jung SM, Zalabany M, *et al.* Carbon-nanotube-embedded hydrogel sheets for engineering cardiac constructs and bioactuators. ACS Nano 2013; 7(3): 2369-80.
[http://dx.doi.org/10.1021/nn305559j] [PMID: 23363247]

[64] Shin SR, Zihlmann C, Akbari M, *et al.* Reduced Graphene Oxide-GelMA Hybrid Hydrogels as Scaffolds for Cardiac Tissue Engineering. Small 2016; 12(27): 3677-89.
[http://dx.doi.org/10.1002/smll.201600178] [PMID: 27254107]

[65] Grosberg A, Alford PW, McCain ML, Parker KK. Ensembles of engineered cardiac tissues for

physiological and pharmacological study: heart on a chip. Lab Chip 2011; 11(24): 4165-73.
[http://dx.doi.org/10.1039/c1lc20557a] [PMID: 22072288]

[66] van Meer PJ, Graham ML, Schuurman HJ. The safety, efficacy and regulatory triangle in drug development: Impact for animal models and the use of animals. Eur J Pharmacol 2015; 759: 3-13.
[http://dx.doi.org/10.1016/j.ejphar.2015.02.055] [PMID: 25818943]

[67] Lalit PA, Hei DJ, Raval AN, Kamp TJ. Induced pluripotent stem cells for post-myocardial infarction repair: remarkable opportunities and challenges. Circ Res 2014; 114(8): 1328-45.
[http://dx.doi.org/10.1161/CIRCRESAHA.114.300556] [PMID: 24723658]

[68] Kurokawa YK, George SC. Tissue engineering the cardiac microenvironment: Multicellular microphysiological systems for drug screening. Adv Drug Deliv Rev 2016; 96: 225-33.
[http://dx.doi.org/10.1016/j.addr.2015.07.004] [PMID: 26212156]

[69] Tzatzalos E, Abilez OJ, Shukla P, Wu JC. Engineered heart tissues and induced pluripotent stem cells: Macro- and microstructures for disease modeling, drug screening, and translational studies. Adv Drug Deliv Rev 2016; 96: 234-44.
[http://dx.doi.org/10.1016/j.addr.2015.09.010] [PMID: 26428619]

[70] Zhou J, Chen J, Sun H, *et al.* Engineering the heart: evaluation of conductive nanomaterials for improving implant integration and cardiac function. Sci Rep 2014; 4: 3733.
[http://dx.doi.org/10.1038/srep03733] [PMID: 24429673]

[71] Chung HJ, Park TG. Surface engineered and drug releasing pre-fabricated scaffolds for tissue engineering. Adv Drug Deliv Rev 2007; 59(4-5): 249-62.
[http://dx.doi.org/10.1016/j.addr.2007.03.015] [PMID: 17482310]

[72] Ye L, Chang Y-H, Xiong Q, *et al.* Cardiac repair in a porcine model of acute myocardial infarction with human induced pluripotent stem cell-derived cardiovascular cells. Cell Stem Cell 2014; 15(6): 750-61.
[http://dx.doi.org/10.1016/j.stem.2014.11.009] [PMID: 25479750]

[73] Li J, Zhu K, Yang S, *et al.* Fibrin patch-based insulin-like growth factor-1 gene-modified stem cell transplantation repairs ischemic myocardium. Exp Biol Med (Maywood) 2015; 240(5): 585-92.
[http://dx.doi.org/10.1177/1535370214556946] [PMID: 25767192]

[74] Jaffe EA. Cell biology of endothelial cells. Hum Pathol 1987; 18(3): 234-9.
[http://dx.doi.org/10.1016/S0046-8177(87)80005-9] [PMID: 3546072]

[75] Chan-Park MB, Shen JY, Cao Y, *et al.* Biomimetic control of vascular smooth muscle cell morphology and phenotype for functional tissue-engineered small-diameter blood vessels. J Biomed Mater Res A 2009; 88(4): 1104-21.
[http://dx.doi.org/10.1002/jbm.a.32318] [PMID: 19097157]

[76] Furchgott RF. Role of endothelium in responses of vascular smooth muscle. Circ Res 1983; 53(5): 557-73.
[http://dx.doi.org/10.1161/01.RES.53.5.557] [PMID: 6313250]

[77] Owens GK. Regulation of differentiation of vascular smooth muscle cells. Physiol Rev 1995; 75(3): 487-517.
[PMID: 7624392]

[78] Naito Y, Shinoka T, Duncan D, *et al.* Vascular tissue engineering: towards the next generation vascular grafts. Adv Drug Deliv Rev 2011; 63(4-5): 312-23.
[http://dx.doi.org/10.1016/j.addr.2011.03.001] [PMID: 21421015]

[79] Nerem RM, Seliktar D. Vascular tissue engineering. Annu Rev Biomed Eng 2001; 3(1): 225-43.
[http://dx.doi.org/10.1146/annurev.bioeng.3.1.225] [PMID: 11447063]

[80] Seifu DG, Purnama A, Mequanint K, Mantovani D. Small-diameter vascular tissue engineering. Nat Rev Cardiol 2013; 10(7): 410-21.
[http://dx.doi.org/10.1038/nrcardio.2013.77] [PMID: 23689702]

[81]   Gui L, Niklason LE. Vascular tissue engineering: building perfusable vasculature for implantation. Curr Opin Chem Eng 2014; 3: 68-74.
[http://dx.doi.org/10.1016/j.coche.2013.11.004] [PMID: 24533306]

[82]   Huang NF, Li S. Mesenchymal stem cells for vascular regeneration. Regen Med 2008; 3(6): 877-92.
[http://dx.doi.org/10.2217/17460751.3.6.877] [PMID: 18947310]

[83]   Hashi CK, Zhu Y, Yang G-Y, *et al.* Antithrombogenic property of bone marrow mesenchymal stem cells in nanofibrous vascular grafts. Proc Natl Acad Sci USA 2007; 104(29): 11915-20.
[http://dx.doi.org/10.1073/pnas.0704581104] [PMID: 17615237]

[84]   Au P, Tam J, Fukumura D, Jain RK. Bone marrow-derived mesenchymal stem cells facilitate engineering of long-lasting functional vasculature. Blood 2008; 111(9): 4551-8.
[http://dx.doi.org/10.1182/blood-2007-10-118273] [PMID: 18256324]

[85]   Sinha S, Hoofnagle MH, Kingston PA, McCanna ME, Owens GK. Transforming growth factor-β1 signaling contributes to development of smooth muscle cells from embryonic stem cells. Am J Physiol Cell Physiol 2004; 287(6): C1560-8.
[http://dx.doi.org/10.1152/ajpcell.00221.2004] [PMID: 15306544]

[86]   Ferreira LS, Gerecht S, Fuller J, Shieh HF, Vunjak-Novakovic G, Langer R. Bioactive hydrogel scaffolds for controllable vascular differentiation of human embryonic stem cells. Biomaterials 2007; 28(17): 2706-17.
[http://dx.doi.org/10.1016/j.biomaterials.2007.01.021] [PMID: 17346788]

[87]   Riha GM, Lin PH, Lumsden AB, Yao Q, Chen C. Review: application of stem cells for vascular tissue engineering. Tissue Eng 2005; 11(9-10): 1535-52.
[http://dx.doi.org/10.1089/ten.2005.11.1535] [PMID: 16259608]

[88]   Lehoux S, Castier Y, Tedgui A. Molecular mechanisms of the vascular responses to haemodynamic forces. J Intern Med 2006; 259(4): 381-92.
[http://dx.doi.org/10.1111/j.1365-2796.2006.01624.x] [PMID: 16594906]

[89]   Traub O, Berk BC. Laminar shear stress: mechanisms by which endothelial cells transduce an atheroprotective force. Arterioscler Thromb Vasc Biol 1998; 18(5): 677-85.
[http://dx.doi.org/10.1161/01.ATV.18.5.677] [PMID: 9598824]

[90]   Dudash LA, Kligman F, Sarett SM, Kottke-Marchant K, Marchant RE. Endothelial cell attachment and shear response on biomimetic polymer-coated vascular grafts. J Biomed Mater Res A 2012; 100(8): 2204-10.
[http://dx.doi.org/10.1002/jbm.a.34119] [PMID: 22623267]

[91]   Smith RJ Jr, Koobatian MT, Shahini A, Swartz DD, Andreadis ST. Capture of endothelial cells under flow using immobilized vascular endothelial growth factor. Biomaterials 2015; 51: 303-12.
[http://dx.doi.org/10.1016/j.biomaterials.2015.02.025] [PMID: 25771020]

[92]   Nieponice A, Maul TM, Cumer JM, Soletti L, Vorp DA. Mechanical stimulation induces morphological and phenotypic changes in bone marrow-derived progenitor cells within a three-dimensional fibrin matrix. J Biomed Mater Res A 2007; 81(3): 523-30.
[http://dx.doi.org/10.1002/jbm.a.31041] [PMID: 17133453]

[93]   Halka AT, Turner NJ, Carter A, *et al.* The effects of stretch on vascular smooth muscle cell phenotype *in vitro.* Cardiovasc Pathol 2008; 17(2): 98-102.
[http://dx.doi.org/10.1016/j.carpath.2007.03.001] [PMID: 18329554]

[94]   Sadoshima J, Jahn L, Takahashi T, Kulik TJ, Izumo S. Molecular characterization of the stretch-induced adaptation of cultured cardiac cells. An *in vitro* model of load-induced cardiac hypertrophy. J Biol Chem 1992; 267(15): 10551-60.
[PMID: 1534087]

[95]   Jeong SI, Kwon JH, Lim JI, *et al.* Mechano-active tissue engineering of vascular smooth muscle using pulsatile perfusion bioreactors and elastic PLCL scaffolds. Biomaterials 2005; 26(12): 1405-11.

[http://dx.doi.org/10.1016/j.biomaterials.2004.04.036] [PMID: 15482828]

[96]   Stickler P, De Visscher G, Mesure L, *et al.* Cyclically stretching developing tissue *in vivo* enhances mechanical strength and organization of vascular grafts. Acta Biomater 2010; 6(7): 2448-56.
[http://dx.doi.org/10.1016/j.actbio.2010.01.041] [PMID: 20123137]

[97]   Wells RG. The role of matrix stiffness in regulating cell behavior. Hepatology 2008; 47(4): 1394-400.
[http://dx.doi.org/10.1002/hep.22193] [PMID: 18307210]

[98]   Lutolf MP, Hubbell JA. Synthetic biomaterials as instructive extracellular microenvironments for morphogenesis in tissue engineering. Nat Biotechnol 2005; 23(1): 47-55.
[http://dx.doi.org/10.1038/nbt1055] [PMID: 15637621]

[99]   Han F, Jia X, Dai D, *et al.* Performance of a multilayered small-diameter vascular scaffold dual-loaded with VEGF and PDGF. Biomaterials 2013; 34(30): 7302-13.
[http://dx.doi.org/10.1016/j.biomaterials.2013.06.006] [PMID: 23830580]

[100] Appel AA, Larson JC, Jiang B, Zhong Z, Anastasio MA, Brey EM. X-ray Phase Contrast Allows Three Dimensional, Quantitative Imaging of Hydrogel Implants. Ann Biomed Eng 2016; 44(3): 773-81.
[http://dx.doi.org/10.1007/s10439-015-1482-5] [PMID: 26487123]

[101] Appel A, Anastasio MA, Brey EM. Potential for imaging engineered tissues with X-ray phase contrast. Tissue Eng Part B Rev 2011; 17(5): 321-30.
[http://dx.doi.org/10.1089/ten.teb.2011.0230] [PMID: 21682604]

[102] Kim K, Wagner WR. Non-invasive and Non-destructive Characterization of Tissue Engineered Constructs Using Ultrasound Imaging Technologies: A Review. Ann Biomed Eng 2016; 44(3): 621-35.
[http://dx.doi.org/10.1007/s10439-015-1495-0] [PMID: 26518412]

[103] Kotecha M, Klatt D, Magin RL. Monitoring cartilage tissue engineering using magnetic resonance spectroscopy, imaging, and elastography. Tissue Eng Part B Rev 2013; 19(6): 470-84.
[http://dx.doi.org/10.1089/ten.teb.2012.0755] [PMID: 23574498]

CHAPTER 4

Liver and Kidney Tissue Engineering

Fatemeh Khatami[1], Mehdi Razavi[2] and Yi-Nan Zhang[3,*]

[1] *Skin Research Center, Shahid Beheshti University of Medical Sciences, Tehran, Iran*

[2] *Department of Radiology, School of Medicine, Stanford University, Palo Alto, California 94304, USA*

[3] *Institute of Biomaterials and Biomedical Engineering, University of Toronto, 164 College Street, Toronto, ON, M5S 3G9, Canada*

Abstract: Liver and kidney are among the most demanded organs for transplantation. There is currently no tissue engineered liver or kidney that has been used as a clinical product. One of the main reasons is due to the structural and functional complexity of these organs. In recent years, several studies have demonstrated significant progress in development of an engineered 3D tissues and decellularization from the original organ. 3D printing bio-microelectromechanical system (BioMEMS) and organ-on-a-chip technology can help to provide a biological *in vitro* model for studying fundamental organ biology, organ disease, toxicology and drug discovery. Despite promising approaches have been used by many research groups within last decades, how to construct a fully functional liver or kidney tissue still reminds a big challenge. In this chapter we will review the most recent advances in liver and kidney tissue engineering, respectively.

Keywords: 3D model, BioMEMS, Decellularization, Kidney, Liver, Microfluidics, Tissue Engineering.

INTRODUCTION

Liver and kidney play important roles in the body. The major role of liver is metabolism, secretion of bile, digestion, energy production and detoxifying substances. The main function of kidney is waste excretion, blood pressure regulation and red blood cell regulation. Both liver and kidney have complex structure, cellular components and functionality which lead to challenges of building up such an organ for clinical transplantation using tissue engineering. Bio-microelectromechanical system (MEMS) technology and organ-on-a-chip approach provide precise control of cellular architecture of liver and kidney with

* **Corresponding author Yi-Nan Zhang:** Institute of Biomaterials and Biomedical Engineering, University of Toronto, Toronto, Canada; Tel: 647-985-0120; E-mail: yinanzhang.zhang@mail.utoronto.ca

Mehdi Razavi (Ed.)
All rights reserved-© 2017 Bentham Science Publishers

vascularization. Other promising approach for liver or kidney tissue engineering involves the use of decellularized donated organ as a natural porous scaffold for cell seeding. The advantages and drawbacks of these approaches will be discussed in this chapter.

STRUCTURE AND FUCNTION OF HUMAN LIVER

The liver is made up of hexagonal prisms with portal and hepatic lobules [1, 2]. Blood flow into the hepatic sinusoids from hepatic arteries and portal veins. The majority of the liver's cell population is composed of hepatocytes, which provide the function of metabolism [3, 4]. Endothelial cells lining in the hepatic sinusoids. Kupffer cells are the resident macrophages as innate immune cell populations in the liver [3]. The liver plays numbers of vital roles such as metabolizing the breakdown products of digestion, and detoxifying substances that are harmful to the body [5]. In addition, regeneration is a unique ability of the liver. After removing some damaged parts, the liver can still continue its function in the body.

TISSUE ENGINEERED LIVER SUBSTITUTES

Cell Based Therapy

Liver disease affects hundreds of thousands of human lives [6]. Liver transplantation is considered as the most effective way to directly prevent mortality. However, donating enough healthy livers is a major problem in the society [7, 8]. Cell based therapies provide an alternative way to build the hepatic microenvironment [9, 10]. More importantly, the primary hepatocytes can be directly taken from human and avoid the differences between species [11]. It has been reported that the human hepatocytes can expand and proliferate with the *in vivo* hepatic stromal environment [12]. However, it still remains challenge to keep the functionality and proliferative ability while isolating and culturing primary hepatocytes *in vitro* since *in vitro* cell culture cannot simply mimic the *in vivo* biological environment [13, 14].

It has been demonstrated that co-culture and using optimized cultural components can improve *in vitro* systems for culturing primary human hepatocytes [2]. The role of microenvironmental signals in governing hepatocellular proliferation has been studied in *in vitro* culture systems. These optimized systems include co-culture to provide specific cell-cell interactions as well as defined concentration of soluble factors and extracellular matrix molecules. However, further constructing a 3D hepatic environment with all cellular components remain a challenge. Kang *et al.* [15] demonstrated a 3D model using transwell membrane system with one layer of primary rat hepatocytes and another layer of primary rat livery sinusoidal endothelial cells. When primary hepatocytes and endothelial cells cultured on the opposite side of transwell membrane, viability and functionality of both cell types were significantly improved. However, primary hepatocytes lost their viability after 5 days without co-culture with the endothelial cells in this 3D system.

Whole Liver Decellularization Method

Since it is difficult to build the complex liver structure, researches have begun concentrating their efforts on providing decellularized liver porous scaffold. Several studies indicated that whole organ decellularization can retain the native liver porous structure and supports liver-specific biological functions [16]. The native scaffolds consist of ECM containing variety of growth factors necessary for cell attachment, differentiation and proliferation [17], and also intact microvasculature which can promote cell distribution and neovascularization [18, 19]. Therefore, developing a decellularization method for removing all cells without adversely affecting its composition, biological activity and mechanical integrity is essential. So far, perfusion decellularization by different concentration of Triton-X100, SDS or a combination of both has been used for liver decellularization [20, 21] and it has been indicated that the detergent based decellularization causes substantial harm to the ECM and vasculature network [22]. Maghsoudlou *et al.* [23] showed that decellularization of liver by detergent-enzymatic technique (EDTA-DET and DET) maintains the ECM components. It has been indicated that the addition of EDTA compared with fresh tissue can increase the collagen content of the scaffold by 300% while elastin content reduces to 40%. These proteins have been shown to be important in porous scaffold microstructure. However, the EDTA shown to deteriorate the vascular network compared to the protocol without EDTA (DET). Maza *et al.* [24] reported that the application of different cell damaging factors including isotonic stress, mechanical stress, and flow shear stress along with enzyme and detergents (T-X100+SDS) perfusion is an effective way in providing a well preserved liver scaffold with no residual cells. Though the combination of detergents resulted in less ECM disruption, the procedure is time consuming and there is also a risk of cytotoxicity by detergents as they penetrate into the thick tissues [25]. Thus, care must be taken to remove residual agents from the ECM after decellularization. With current advances in decellularization protocols, whole-liver scaffold with minimal loss of ECM components will eventually pave the way for liver tissue engineering.

Bioreactor for Liver Culture

In vitro models that perform the complication of *in vivo* tissue and organ behaviors in a scalable and easy-to-use format are necessary [26]. Accordingly, current research has mostly been focused on the development of new culture techniques or new immortalized hepatic cell lines. A main problem to the generation of functional substitutes is the restricted understanding of the role of specific physicochemical parameters on tissue development. Bioreactors provide controlled environmental conditions to improve the quality of tissue and to investigate the liver cell metabolism. Bioreactors are widely utilized at any step of the assembly of a tissue engineered liver construct. The cell-seeded porous scaffolds are then cultured in bioreactors under tightly controlled and closely monitored environmental conditions to provide cells with biochemical and physical cues that should encourage cell reorganization into liver-like aggregates and differentiation to create the construct functionally equivalent to liver tissue [27]. Several bioreactor systems that capture some aspects of *in vivo* physiology have been developed for liver cell culture, ranging from flat-plate or matrix-sandwiched monolayer designs [28, 29] to three-dimensional (3D) perfusion culture employing membranes [30], beads [31], or polymeric meshes [32] to retain cells *in situ*. The formation of hepatic structures in a perfused environment seems to offer some of the physiological properties, forces, and dimensions of native *in vivo* liver tissue. Powers *et al.* [33] indicated that this indeed results in an enhanced and sustained functional response. Domansky *et al.* [26] also developed a bioreactor that fosters maintenance of 3D tissue cultures under constant perfusion and multiple bioreactors have been integrated into an array in a multiwell plate format. The designed perfused multiwell can be employed not only for the culture of liver cells, it can also be utilized for perfusion culture of other high metabolically active cell types such as kidney, heart, or brain cells. Hence, these reactor systems provide a promising platform for *in vitro* study of *in vivo* hepatic phenomena [33]. Though, much work is still need to be done towards the understanding and the modeling of transport phenomena in the proposed bioreactors [27].

BioMEMS Scaffolding Tissue Engineering

To date, there is currently no tissue-engineered liver that has been used as a clinical product [34]. One of the most important reasons is due to the complexity of liver structure and function. In order to counter this barrier, MEMS technology allows the precise control of cellular microenvironments, which can result in vascularization [35, 36]. Hydrogel, which serves as porous scaffold, can provide the biocompatible and biodegradable hepatic microenvironment for cell seeding. Endothelial cells are essential for vascularization and built into the microfluidic channels while culturing with the hepatocytes in the device. In addition, this device can provide nutrients and waste production, dynamic fluids and mechanical contraction, which closely mimic the *in vivo* hepatic sinusoid microarchitecture using an organ-on-a-chip approach [37 - 39]. Nakao *et al.* [40] reported that they are able to keep the metabolic activity of primary human hepatocytes and induce the formation of bile canaliculi in the device. It indicates the primary hepatocytes have the similar function compared to the *in vivo* biological environment. To improve the complexity of the architecture and multi-function of an artificial organ, 3D printing BioMEMS technology allows to build into multi-cellular components, multi-channels and stimulating factors using a bottom-up approach [38]. Further, this microfluidic device can also be used for drug testing, such as toxicity and cell metabolic studies.

STRUCTURE AND FUNCTION OF HUMAN KIDNEY

Kidney has a complex structure with more than 30 different cell types. Kidney is divided into two major structures: renal cortex and renal medulla. It contains nephrons and renal tubules that pass from the cortex deep into the medullary pyramids [41]. It also consists of numerous arterioles and capillaries to filter the bloodstream and excrete the waste through urine [42]. Kidney extracellular matrix is composed of variety of proteins, growth factors and glycosaminoglycans which provide specific cues for modulation of cell behavior and function [43].

STRATEGIES FOR TISSUE ENGINEERING OF KIDNEY

End-stage renal disease (ESRD) is one of the life-threatening disease all over the world. The two treatments for ESRD are dialysis or a kidney transplant. Transplantation is much more satisfactory because it provides the full range of renal function [44]. However, the availability of immunologically match donor is the main problem for transplantation. To address this shortage, several strategies are being used to regenerate a whole functional human kidney. These approaches are based on stem cell therapy with a supporting porous scaffold which is ideally similar to the ECM of kidney. Due to the complex architecture and lineage specification of kidney, it is one of the most challenging organs to engineer. There are several potential approaches for kidney tissue engineering such as repopulation of a decellularized postmortem kidney with fresh stem cells, *in situ* regeneration by recruiting endogenous stem cells or developing a completely new construct by 3D printing technology [45, 46].

Kidney Decellularization and Bioreactor for Kidney Culture

Decellularization is one of the most promising method for kidney tissue engineering. The advantage of this biological scaffold is proper geometric locations of ECM molecules and intact vascular conduits. Moreover, the presence of biological molecules such as growth factors and cytokines can be beneficial for cell growth and functionality [47, 48]. There are several protocols for organ decellularization which are mostly based on the use of detergents (*e.g.* Triton X-100, SDS) or enzymatic detachment of cells by trypsin to solubilize and wash out cellular components. The resulting scaffold must be evaluated for structural integrity, retention of biological agents like growth factors as well as appropriate cell removal [49]. In this respect, Caralt *et al.* [48] reported that perfusion of 0.1% SDS following Triton leads to superior cell removal and retention of growth factors while trypsin-based decellularization is not effective for kidney and causes growth factor loss. ECM-bound growth factors (*e.g.* bFGF and VEGF) promote cell proliferation, differentiation and angiogenesis. In addition, it has been shown that kidney decellularization by detergent perfusion and subsequently seeding the porous scaffold with epithelial and endothelial cells can result in a functional kidney construct *in vitro* and with an excretory function *in vivo* after transplantation [43]. Therefore, detergent based decellularization is an effective strategy to remove cell debris without disruption of vascular, glomerular and tubular ultrastructure [50, 51].

Bioreactor systems enables the perfusion of kidney decellularization *via* the vasculature [43, 52]. Generally, renal artery and ureter are the main root of cell delivery for kidney [43]. Caralt *et al.* [48] recently designed a perfusion-based bioreactor to infuse immortalized human renal cortical tubular epithelial (RCTE) cells through the renal artery. The outcomes showed 50% recellularization after a day and the tubule formation with increased metabolic activity. Song *et al.* [43] reported to infuse the porous scaffold using neonatal rat kidney cells through the ureter and further culture the entire kidney in a bioreactor. This approach provide a strategy for kidney cell seeding using bioreactors.

De Novo Kidney Generation

De novo kidney regeneration is basically referred to the use of 3D printing technology to design an appropriate 3D scaffold or relies on cells' own capacity to organize themselves, as they do in embryonic development. In 3D printing approach organs with complicated vascular systems, such as the kidney and liver, could be replicated. Subsequently, the autologous cells and biological materials will be deposited on the scaffold [46, 53]. However, there has not yet been any published report about the success of this technique for generation of functional kidney. Another approach relies on the application of embryonic stem cells without any external scaffold. One of the first established methods was based on the dissociation of embryonic kidneys into single-cells suspension and reaggregation in the presence of drugs to form organotropic renal constructions [54]. Recent progress in culture systems has made it possible to develop intact fetal kidney with distinct cortex and medulla. Furthermore, this engineered kidney becomes vascularized after transferring to an *in vivo* condition (chick egg chorioallantoic membrane or adult rat). Despite the advancement of this approach in engineering of organs, they are far from clinical application. For instance, they are very small and flat (lack of complete 3D structure of native kidney) and from murine fetal stem cells but not human cells [55, 56].

Kidney-on-a-Chip and 3D Bioprinting

Many researches have been carried out for designing kidney on a chip. One of the initial designs for recapitulating functions of the renal tubule comprised of two compartments which mimic the urinary lumen and interstitial space. The results showed that the fluidic sheer stress in the upper chamber improved the cytoskeletal reorganization and cell polarization [57]. Another strategy is based on fabricating cylindrical microchannel and coating PDMS microchannels with glass. This approach enabled uniform seeding of HK-2 cells on the cylindrical walls which provides a suitable model to investigate kidney-related diseases [58]. Besides, Ng *et al.* [59] introduced a bioartificial kidney device comprising tubular hollow-fiber membrane embedded within a multi-compartment microfluidic system which enables the fluid independently to flow inside and outside the tubule. The results indicated successful monolayer proximal tubule epithelial cell attachment along the luminal surface coated with fibrin as well as active transportation through the tubule [59]. Kidney-on-a-chip has various potential applications for physiological experiments, nephrotoxic drugs, measuring kidney injury and so forth. 3D printing technology has been applied to construct the renal tumor model [60]. Mu *et al.* [61] fabricated 3D vascular network construction using microfluidic method to replicate passive diffusion in a nephron. 3D printing of whole kidney organ is still challenging and in order to achieve a functional 3D printed kidney, some major barriers including appropriate biomaterials, correctly differentiated cell types and the mechanical stability of the constructs, with the aim of mimicking a native kidney tissue-like microenvironment should be addressed.

CONCLUDING REMARKS

The clinical application of engineered organs will demand the optimization of several specific variables for the organ in question. Development of transplantable 3D-printed liver and kidney requires recapitulation of tissue architecture and function to precisely simulate the *in vivo* organ conditions. BioMEMS technology and co-culture systems can enhance our understanding of fundamental organ biology. Synthetic and metabolic functions of liver and kidney stem cells can be optimized on an appropriate porous scaffold which presents essential cues for cell proliferation and differentiation. Currently, decellularized whole liver and kidney scaffolds have shown promising outcomes for whole organ regeneration, not to mention there are still many hurdles to be overcome.

ABBREVIATION

iPS	induced pluripotent stem
3D	three-dimensional
PRHs	primary rat hepatocytes
LSECs	livery sinusoidal endothelial cells
BAECs	bovine aortic endothelial cells
ECM	extracellular matrix
MEMS	microelectromechanical system
SDS	sodium dodecyl sulfate
bFGF	basic fibroblast growth factor
ECM	extracellular matrix
VEGF	vascular endothelial growth factor
ESRD	end-stage renal disease
hNPCs	human nephron progenitor cells
hESC	human embryonic stem cell
RCTE	immortalized human renal cortical tubular epithelial cell

CONFLICT OF INTEREST

The authors declare no conflict of interest, financial or otherwise.

ACKNOWLEDGEMENTS

Declared none.

REFERENCES

[1]   Juza RM, Pauli EM. Clinical and surgical anatomy of the liver: a review for clinicians. Clin Anat 2014; 27(5): 764-9.
[http://dx.doi.org/10.1002/ca.22350] [PMID: 24453062]

[2]   Bhatia SN, Underhill GH, Zaret KS, Fox IJ. Cell and tissue engineering for liver disease. Sci Transl Med 2014; 6(245): 245sr2.
[http://dx.doi.org/10.1126/scitranslmed.3005975] [PMID: 25031271]

[3]   Zhang YN, Poon W, Tavares AJ, McGilvray ID, Chan WC. Nanoparticle-liver interactions: Cellular uptake and hepatobiliary elimination. J Control Release 2016; 240: 332-48.
[http://dx.doi.org/10.1016/j.jconrel.2016.01.020] [PMID: 26774224]

[4]   Rambhatla L, Chiu CP, Kundu P, Peng Y, Carpenter MK. Generation of hepatocyte-like cells from human embryonic stem cells. Cell Transplant 2003; 12(1): 1-11.
[http://dx.doi.org/10.3727/000000003783985179] [PMID: 12693659]

[5]   Hoekstra LT, de Graaf W, Nibourg GA, *et al.* Physiological and biochemical basis of clinical liver function tests: a review. Ann Surg 2013; 257(1): 27-36.
[http://dx.doi.org/10.1097/SLA.0b013e31825d5d47] [PMID: 22836216]

[6]   Shepard CW, Finelli L, Alter MJ. Global epidemiology of hepatitis C virus infection. Lancet Infect Dis 2005; 5(9): 558-67.
[http://dx.doi.org/10.1016/S1473-3099(05)70216-4] [PMID: 16122679]

[7]   Lin HM, Kauffman HM, McBride MA, *et al.* Center-specific graft and patient survival rates: 1997 United Network for Organ Sharing (UNOS) report. JAMA 1998; 280(13): 1153-60.
[http://dx.doi.org/10.1001/jama.280.13.1153] [PMID: 9777815]

[8]   Khademhosseini A, Vacanti JP, Langer R. Progress in tissue engineering. Sci Am 2009; 300(5): 64-71.
[http://dx.doi.org/10.1038/scientificamerican0509-64] [PMID: 19438051]

[9]   Fiegel HC, Kaufmann PM, Bruns H, *et al.* Hepatic tissue engineering: from transplantation to customized cell-based liver directed therapies from the laboratory. J Cell Mol Med 2008; 12(1): 56-66.
[http://dx.doi.org/10.1111/j.1582-4934.2007.00162.x] [PMID: 18021311]

[10]   Shan J, Schwartz RE, Ross NT, *et al.* Identification of small molecules for human hepatocyte expansion and iPS differentiation. Nat Chem Biol 2013; 9(8): 514-20.
[http://dx.doi.org/10.1038/nchembio.1270] [PMID: 23728495]

[11]   Gibbs RA, Weinstock GM, Metzker ML, *et al.* Genome sequence of the Brown Norway rat yields insights into mammalian evolution. Nature 2004; 428(6982): 493-521.
[http://dx.doi.org/10.1038/nature02426] [PMID: 15057822]

[12]   Michalopoulos GK, DeFrances MC. Liver regeneration. Science 1997; 276(5309): 60-6.
[http://dx.doi.org/10.1126/science.276.5309.60] [PMID: 9082986]

[13]   Mitaka T. The current status of primary hepatocyte culture. Int J Exp Pathol 1998; 79(6): 393-409.
[http://dx.doi.org/10.1046/j.1365-2613.1998.00083.x] [PMID: 10319020]

[14]   Runge D, Michalopoulos GK, Strom SC, Runge DM. Recent advances in human hepatocyte culture systems. Biochem Biophys Res Commun 2000; 274(1): 1-3.
[http://dx.doi.org/10.1006/bbrc.2000.2912] [PMID: 10903886]

[15]   Kang YB, Rawat S, Cirillo J, Bouchard M, Noh HM. Layered long-term co-culture of hepatocytes and endothelial cells on a transwell membrane: toward engineering the liver sinusoid. Biofabrication 2013; 5(4): 045008.
[http://dx.doi.org/10.1088/1758-5082/5/4/045008] [PMID: 24280542]

[16]   Wang Y, Cui CB, Yamauchi M, *et al.* Lineage restriction of human hepatic stem cells to mature fates is made efficient by tissue-specific biomatrix scaffolds. Hepatology 2011; 53(1): 293-305.
[http://dx.doi.org/10.1002/hep.24012] [PMID: 21254177]

[17]   Badylak SF, Taylor D, Uygun K. Whole-organ tissue engineering: decellularization and recellularization of three-dimensional matrix scaffolds. Annu Rev Biomed Eng 2011; 13: 27-53.
[http://dx.doi.org/10.1146/annurev-bioeng-071910-124743] [PMID: 21417722]

[18]   Moon JJ, West JL. Vascularization of engineered tissues: approaches to promote angio-genesis in biomaterials. Curr Top Med Chem 2008; 8(4): 300-10.
[http://dx.doi.org/10.2174/156802608783790983] [PMID: 18393893]

[19]   Sabetkish S, Kajbafzadeh A-M, Sabetkish N. Whole-organ tissue engineering: Decellularization and recellularization of three-dimensional matrix liver scaffolds. J Biomed Mater Res Part A 2014. 00A:000–000

[20]   Uygun BE, Soto-Gutierrez A, Yagi H, *et al.* Organ reengineering through development of a transplantable recellularized liver graft using decellularized liver matrix. Nat Med 2010; 16(7): 814-20.
[http://dx.doi.org/10.1038/nm.2170] [PMID: 20543851]

[21]   Kajbafzadeh A, Javan-farazmand N, Monajemzadeh M, Baghayee A. Determining the optimal decellularization and sterilization protocol for preparing a tissue scaffold of a human-sized liver tissue 2013; 19: 642-51.

[22] Shirakigawa N, Ijima H, Takei T. Decellularized liver as a practical scaffold with a vascular network template for liver tissue engineering. J Biosci Bioeng 2012; 114(5): 546-51.
[http://dx.doi.org/10.1016/j.jbiosc.2012.05.022] [PMID: 22717723]

[23] Maghsoudlou P, Georgiades F, Smith H, *et al.* Optimization of liver decellularization maintains extracellular matrix micro-architecture and composition predisposing to effective cell seeding. PLoS One 2016; 11(5): e0155324.
[http://dx.doi.org/10.1371/journal.pone.0155324] [PMID: 27159223]

[24] Mazza G, Rombouts K, Rennie Hall A, *et al.* Decellularized human liver as a natural 3D-scaffold for liver bioengineering and transplantation. Sci Rep 2015; 5: 13079.
[http://dx.doi.org/10.1038/srep13079] [PMID: 26248878]

[25] Cebotari S, Tudorache I, Jaekel T, *et al.* Detergent decellularization of heart valves for tissue engineering: toxicological effects of residual detergents on human endothelial cells. Artif Organs 2010; 34(3): 206-10.
[http://dx.doi.org/10.1111/j.1525-1594.2009.00796.x] [PMID: 20447045]

[26] Domansky K, Inman W, Serdy J, Dash A, Lim MH, Griffith LG. Perfused multiwell plate for 3D liver tissue engineering. Lab Chip 2010; 10(1): 51-8.
[http://dx.doi.org/10.1039/B913221J] [PMID: 20024050]

[27] Catapano G, Gerlach JC. Bioreactors for Liver Tissue Engineering. Top Tissue Eng 2007; 3: 1-42.

[28] Bader A, Frühauf N, Zech K, *et al.* Development of a small-scale bioreactor for drug metabolism studies maintaining hepatospecific functions. Xenobiotica 1998; 28(9): 815-25.
[http://dx.doi.org/10.1080/004982598239074] [PMID: 9764925]

[29] Ledezma GA, Folch A, Bhatia SN, Balis UJ, Yarmush ML, Toner M. Numerical model of fluid flow and oxygen transport in a radial-flow microchannel containing hepatocytes. J Biomech Eng 1999; 121(1): 58-64.
[http://dx.doi.org/10.1115/1.2798043] [PMID: 10080090]

[30] Wu FJ, Peshwa MV, Cerra FB, Hu WS. Entrapment of hepatocyte spheroids in a hollow fiber bioreactor as a potential bioartificial liver. Tissue Eng 1995; 1(1): 29-40.
[http://dx.doi.org/10.1089/ten.1995.1.29] [PMID: 19877913]

[31] Michalopoulos GK, Bowen WC, Zajac VF, *et al.* Morphogenetic events in mixed cultures of rat hepatocytes and nonparenchymal cells maintained in biological matrices in the presence of hepatocyte growth factor and epidermal growth factor. Hepatology 1999; 29(1): 90-100.
[http://dx.doi.org/10.1002/hep.510290149] [PMID: 9862855]

[32] Kaihara S, Kim S, Kim BS, Mooney DJ, Tanaka K, Vacanti JP. Survival and function of rat hepatocytes cocultured with nonparenchymal cells or sinusoidal endothelial cells on biodegradable polymers under flow conditions. J Pediatr Surg 2000; 35(9): 1287-90.
[http://dx.doi.org/10.1053/jpsu.2000.9298] [PMID: 10999680]

[33] Powers MJ, Janigian DM, Wack KE, Baker CS, Beer Stolz D, Griffith LG. Functional behavior of primary rat liver cells in a three-dimensional perfused microarray bioreactor. Tissue Eng 2002; 8(3): 499-513.
[http://dx.doi.org/10.1089/107632702760184745] [PMID: 12167234]

[34] Sudo R. Multiscale tissue engineering for liver reconstruction. Organogenesis 2014; 10(2): 216-24.
[http://dx.doi.org/10.4161/org.27968] [PMID: 24500493]

[35] Khademhosseini A, Langer R, Borenstein J, Vacanti JP. Microscale technologies for tissue engineering and biology. Proc Natl Acad Sci USA 2006; 103(8): 2480-7.
[http://dx.doi.org/10.1073/pnas.0507681102] [PMID: 16477028]

[36] Borenstein JT, Terai H, King KR, Weinberg EJ, Kaazempur-Mofrad MR, Vacanti JP. Microfabrication technology for vascularized tissue engineering. Biomed Microdevices 2002; 4(3): 167-75.

[http://dx.doi.org/10.1023/A:1016040212127]

[37] Goral VN, Hsieh YC, Petzold ON, Clark JS, Yuen PK, Faris RA. Perfusion-based microfluidic device for three-dimensional dynamic primary human hepatocyte cell culture in the absence of biological or synthetic matrices or coagulants. Lab Chip 2010; 10(24): 3380-6.
[http://dx.doi.org/10.1039/c0lc00135j] [PMID: 21060907]

[38] Huh D, Hamilton GA, Ingber DE. From 3D cell culture to organs-on-chips. Trends Cell Biol 2011; 21(12): 745-54.
[http://dx.doi.org/10.1016/j.tcb.2011.09.005] [PMID: 22033488]

[39] Bhatia SN, Ingber DE. Microfluidic organs-on-chips. Nat Biotechnol 2014; 32(8): 760-72.
[http://dx.doi.org/10.1038/nbt.2989] [PMID: 25093883]

[40] Nakao Y, Kimura H, Sakai Y, Fujii T. Bile canaliculi formation by aligning rat primary hepatocytes in a microfluidic device. Biomicrofluidics 2011; 5(2): 22212.
[http://dx.doi.org/10.1063/1.3580753] [PMID: 21799718]

[41] Clapp WL. Renal Anatomy. In: Zhou XJ, Laszik Z, Nadasdy T, D'Agati VD, Silva FG, Eds. Silva's Diagnostic Renal Pathology. New York: Cambridge University Press 2009.

[42] Guimaraes-Souza N, Soler R, Yoo JJ. Regenerative medicine of the kidney. Biomaterials and tissue engineering in urology. 1st ed. FL: CRC Press LLC 2009; pp. 502-17.
[http://dx.doi.org/10.1533/9781845696375.3.502]

[43] Song JJ, Guyette JP, Gilpin SE, Gonzalez G, Vacanti JP, Ott HC. Regeneration and experimental orthotopic transplantation of a bioengineered kidney. Nat Med 2013; 19(5): 646-51.
[http://dx.doi.org/10.1038/nm.3154] [PMID: 23584091]

[44] Wolfe RA, Ashby VB, Milford EL, *et al.* Comparison of mortality in all patients on dialysis, patients on dialysis awaiting transplantation, and recipients of a first cadaveric transplant. N Engl J Med 1999; 341(23): 1725-30.
[http://dx.doi.org/10.1056/NEJM199912023412303] [PMID: 10580071]

[45] Batchelder CA, Martinez ML, Tarantal AF. Natural Scaffolds for Renal Differentiation of Human Embryonic Stem Cells for Kidney Tissue Engineering. PLoS One 2015; 10(12): e0143849.
[http://dx.doi.org/10.1371/journal.pone.0143849] [PMID: 26645109]

[46] Davies JA, Chang C-H, Lawrence ML, Mills CG, Mullins JJ. Engineered kidneys: principles progress and prospects. Adv Regen Biol 2014; 1: 24990.
[http://dx.doi.org/10.3402/arb.v1.24990]

[47] Yu YL, Shao YK, Ding YQ, *et al.* Decellularized kidney scaffold-mediated renal regeneration. Biomaterials 2014; 35(25): 6822-8.
[http://dx.doi.org/10.1016/j.biomaterials.2014.04.074] [PMID: 24855960]

[48] Caralt M, Uzarski JS, Iacob S, *et al.* Optimization and critical evaluation of decellularization strategies to develop renal extracellular matrix scaffolds as biological templates for organ engineering and transplantation. Am J Transplant 2015; 15(1): 64-75.
[http://dx.doi.org/10.1111/ajt.12999] [PMID: 25403742]

[49] Uzarski JS, Xia Y, Belmonte JC, Wertheim JA. New strategies in kidney regeneration and tissue engineering. Curr Opin Nephrol Hypertens 2014; 23(4): 399-405.
[http://dx.doi.org/10.1097/01.mnh.0000447019.66970.ea] [PMID: 24848937]

[50] Song JJ, Ott HC. Organ engineering based on decellularized matrix scaffolds. Trends Mol Med 2011; 17(8): 424-32.
[http://dx.doi.org/10.1016/j.molmed.2011.03.005] [PMID: 21514224]

[51] Ott HC, Clippinger B, Conrad C, *et al.* Regeneration and orthotopic transplantation of a bioartificial lung. Nat Med 2010; 16(8): 927-33.
[http://dx.doi.org/10.1038/nm.2193] [PMID: 20628374]

[52]   Bijonowski BM, Miller WM, Wertheim JA. Bioreactor design for perfusion-based, highly-vascularized organ regeneration. Curr Opin Chem Eng 2013; 2(1): 32-40.
[http://dx.doi.org/10.1016/j.coche.2012.12.001] [PMID: 23542907]

[53]   Kasyanov V, Brakke K, Vilbrandt T, *et al.* Toward organ printing: design characteristics, virtual modelling and physical prototyping vascular segments of kidney arterial tree. Virtual Phys Prototyp 2011; 6: 197-213.
[http://dx.doi.org/10.1080/17452759.2011.631738]

[54]   Unbekandt M, Davies JA. Dissociation of embryonic kidneys followed by reaggregation allows the formation of renal tissues. Kidney Int 2010; 77(5): 407-16.
[http://dx.doi.org/10.1038/ki.2009.482] [PMID: 20016472]

[55]   Davies JA, Chang CH. Engineering kidneys from simple cell suspensions: an exercise in self-organization. Pediatr Nephrol 2014; 29(4): 519-24.
[http://dx.doi.org/10.1007/s00467-013-2579-4] [PMID: 23989397]

[56]   Xinaris C, Benedetti V, Rizzo P, *et al. In vivo* maturation of functional renal organoids formed from embryonic cell suspensions. J Am Soc Nephrol 2012; 23(11): 1857-68.
[http://dx.doi.org/10.1681/ASN.2012050505] [PMID: 23085631]

[57]   Jang KJ, Suh KY. A multi-layer microfluidic device for efficient culture and analysis of renal tubular cells. Lab Chip 2010; 10(1): 36-42.
[http://dx.doi.org/10.1039/B907515A] [PMID: 20024048]

[58]   Wei Z, Amponsah PK, Al-Shatti M, Nie Z, Bandyopadhyay BC. Engineering of polarized tubular structures in a microfluidic device to study calcium phosphate stone formation. Lab Chip 2012; 12(20): 4037-40.
[http://dx.doi.org/10.1039/c2lc40801e] [PMID: 22960772]

[59]   Ng CP, Zhuang Y, Lin AWH, Teo JCM. A fibrin-based tissueengineered renal proximal tubule for bioartificial kidney devices development, characterization and *in vitro* transport study. Int J Tissue Eng 2013; 2013: 10.

[60]   Bernhard JC, Isotani S, Matsugasumi T, *et al.* Personalized 3D printed model of kidney and tumor anatomy: a useful tool for patient education. World J Urol 2016; 34(3): 337-45.
[http://dx.doi.org/10.1007/s00345-015-1632-2] [PMID: 26162845]

[61]   Mu X, Zheng W, Xiao L, Zhang W, Jiang X. Engineering a 3D vascular network in hydrogel for mimicking a nephron. Lab Chip 2013; 13(8): 1612-8.
[http://dx.doi.org/10.1039/c3lc41342j] [PMID: 23455642]

CHAPTER 5

Skin Substitutes: Current Applications and Challenges

Fatemeh Khatami[1], Reza M. Robati[1], Monireh Torabi-Rahvar[2] and Wanting Niu[3,4,*]

[1] *Skin Research Center, Shahid Beheshti University of Medical Sciences, Tehran, Iran*

[2] *Liver and Pancreatobiliary Diseases Research Center, Digestive Disease Research Institute, Tehran University of Medical Sciences, Tehran, Iran*

[3] *VA Boston Healthcare System, Boston, USA*

[4] *Department of Orthopedic Surgery, Brigham & Women's Hospital, Harvard Medical School, Boston, USA*

Abstract: Skin substitutes as an alternative to skin grafts provide immediate protective barrier against the environment and have critical medical applications to patients with extensive burns. Their initial role is to facilitate repair and reconstruction of skin layers and in more advanced approaches to maintain normal functionality and aesthetics of normal skin. Depending on their design, skin substitutes can act as temporary covers or permanent skin replacements with the main goals of reducing the need for donor sites and to decrease the risk of infection. Moreover, they minimize scarring and also can facilitate angiogenesis. However, current skin substitutes do not restore the normal skin anatomy and they lack skin appendages like sweat glands, hair follicles as well as immune cells. Recent progresses in stem cell biology and engineering techniques hold promise for simulating more advanced skin equivalents with the potential to serve as a drug screening and immune competent models.

Keywords: Dermo-epidermal Substitutes, Skin Equivalent, Skin Graft, Skin Stem Cells, Tissue Engineering, Wound Healing.

INTRODUCTION

Skin is the largest organ in the body which serves as an anatomical barrier to the external environment. This multilayer three-dimensional construct consists of epidermis, dermis and hypodermis made up of cells from all embryological layers. The outer epidermal layer dominated with keratinocytes as the major cell type (90-95%), melanocytes and Langerhans cells (dendritic cells that contribute to the

* **Corresponding author Wanting Niu:** VA Boston Healthcare System; Brigham and Women's Hospital, Harvard Medical School, Boston, USA; Tel: +1-617-637-6609; E-mail: wniu@partners.org

Mehdi Razavi (Ed.)

All rights reserved-© 2017 Bentham Science Publishers

immunological responses of the skin) with less density [1]. The dermis is composed of the papillary dermis and the reticular dermis with fibroblasts as the prominent cell type, which involves constant secretion of collagen, proteoglycan and glycosaminoglycan matrix [2]. These extracellular components are responsible for the resilience and mechanical properties of the skin [3] (Fig. 1). Skin defects caused by trauma, burns, vascular disease, chronic wounds and cancer represent significant clinical problems [4]. The conventional therapies are mostly based on skin grafting techniques to provide xenograft, allograft or autograft split-thickness skin substitutes. Although autograft substitutes are used frequently, there are limitations such as donor site shortage or scar formation after transplantation. For instance, poor healing is observed when wounds extend into the dermis and the subsequent scar tissue lacks elasticity and strength of a normal dermis. Allografts and xenografts made of pig skin serve as temporary dressing to protect the wound since they get rejected by the host immune system after a week [5]. Therefore, tissue engineering approaches have gained much attention over the last decades.

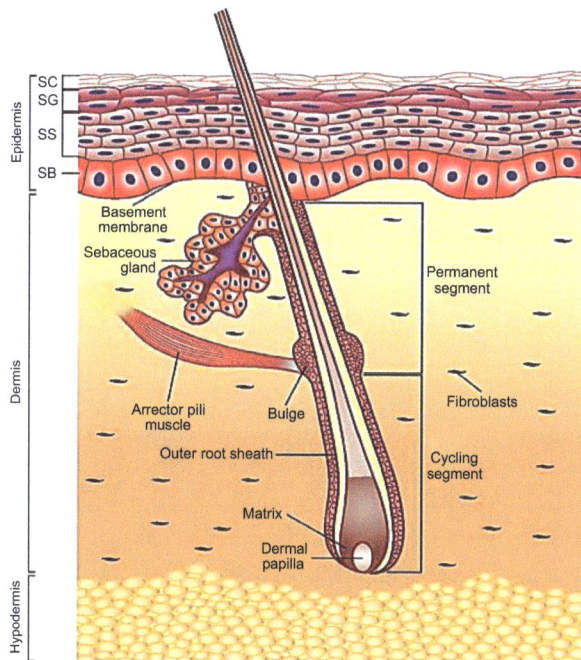

Fig. (1). Skin is comprised of three layers: epidermis, dermis and hypodermis. Epidermis is a stratified squamous epithelium that is divided into four layers of stratum corneum (SC), stratum granulosum (SG), stratum spinosum (SS), and stratum basale (SB). Outer root sheath of the hair follicle is contiguous with the basal epidermal layer. The dermis has a rich vascular system which provides all kinds of cell with nutrient. Fibroblasts are major cell type of the dermis responsible for the synthesis of extracellular matrix. Stem cell niches include the basal epidermal layer, base of sebaceous gland, hair follicle bulge, dermal papillae, and dermis. The hypodermis consists primarily of adipose tissue and serves as an insulator [6].

The primary goal of skin tissue engineering is the generation of implantable substitutes to promote wound healing process [7]. An ideal substitute has to be non-toxic, non-immunogenic, and biodegradable with appropriate mechanical stability. The application of such alternatives prevents fluid loss and heat and protects the wound from infection [8]. In this chapter we have an overview of current tissue engineered-based methods for epidermal, dermal and dermo-epidermal (composite) equivalents.

TISSUE-ENGINEERED SKIN SUBSTITUTES

Skin was the first engineered tissue in the laboratory which has ever reached the marketplace, initially with the development of biodegradable biomaterials and subsequently the significant progress of cell culture techniques. In general there are two major approaches including matrix-based products and cell-based products as alternatives for traditional skin grafts [9] (Table **1**).

Matrix-based Approaches

Acellular substitutes are mainly produced from synthetic polymers or natural biomaterials. The most prevalent synthetic polymers are polyglycolic acid (PGA), polycaprolactone (PCL), polylactic-*co*-glycolic acid (PLGA) and polyethylene terephthalate (PET) [10]. Having consistency between samples, low cost and appropriate physical stability are the advantages of synthetic polymers compared to natural biomaterials.

However, the low yield of cellular attachment and tissue compatibility are the main impediments for clinical applications. In contrast, natural biomaterials such as collagen, elastin, glycosaminoglycan (GAGs), and fibronectin represent high yield of cell attachment, low toxicity and low inflammatory response, albeit, with a poor mechanical strength and stability which necessitate the use of cross-linking agents to improve durability of the matrix. Cross-linked biomaterials are usually being used where the stability of the matrix is more important than cell infiltration (tendon, hernia) and because of associated cell toxicity, cross-linked biomaterials are less suitable for wound healing purposes [11, 12]. The combination of natural biomaterial and synthetic polymers sometimes is used to utilize the advantages of both material components.

Biobrane (Smith & Nephew, Hull, UK) is a dermo-epidermal analog made up of ultrathin silicone film and nylon filament coated with type I collagen. This synthetic sheet acts as a temporary wound dressing. As Biobrane will never get incorporated, it is considered as a wound dressing rather than as a skin substitute.

Table 1. Commercial Skin Substitutes.

Human Skin Substitutes	Commercial Name	Cell Source	Matrix Composition
Epidermal Substitutes	Epicel	Autologous Keratinocytes	Silicone membrane
	Biobrane	Cell-free Synthetic Sheet	A fine nylon mesh cross-linked with porcine dermal collagen
	Epidex	Autologous Keratinocytes Cell-sheet	None
	Laserskin	Autologous Keratinocytes	benzyl esterified hyaluronic acid derivative with ordered laser-perforated microholes
	MySkin	Autologous Keratinocytes	silicon support treated with plasma polymer
Dermal Substitutes	Integra	Cell-free Synthetic Matrix	A silicone layer +collagen+ chondroitin-6-sulfate
	Alloderm	Cell-free Allograft Matrix	Acellular cadaver skin
	Permacol	Cell-free Xenograft Matrix	A porcine-derived acellular dermal matrix
	Dermagraft	Allogenic Fibroblasts	Absorbable PLGA scaffold
	ICX-SKN	Allogenic Fibroblasts	Fibrin matrix
Composite Substitutes	Apligraft	Allogenic Keratinocytes and Fibroblasts	Bovine type I collagen matrix
	OrCel	Allogenic Keratinocytes and Fibroblasts	Non-porous bovine collagen+ porous cross-linked bovine collagen
	PermaDerm	Cell-free Matrix	Collagen-GAG matrix
	Hyalograft-3D	Autologous Fibroblasts	Esterified hyaluronic acid fibers + silicone membrane

There is no significant clinical evidence to prioritize this product over other available substitutes [13].

The first commercially available matrix, Integra (Integra LifeSciences Corporation, Plainsboro, NJ, USA), consists of a cross-linked matrix of bovine collagen and chondroitin-6-sulfate covered with a silicon layer as a temporary epidermis [14]. Alloderm (LifeCell Corporation, The Woodlands, TX, USA) is another dermal skin substitute, derived from acellular cadaver skin. These substitutes are used for both wound healing and tissue reconstruction and the matrix undergoes degradation after host's cells infiltration and proliferation. However, the epidermal layer must eventually be applied to allow re-epithelialization [15]. Animal-derived acellular matrices are also available such as

Permacol (Tissue Science Laboratories, Hampshire, UK) and EZ-derm (Mölnlycke Health Care AB, Gothenburg, Sweden). These products are derived from porcine and undergo chemical cross-linking which makes them less suitable for wound healing purposes. Besides, clinical outcomes have been unconvincing [15, 16].

More recent advances in skin substitutes involve pre-population of matrix with different cell types.

Cell-based Approaches: Human Skin Equivalent

The first landmarks in human skin equivalent (HSE) began with the availability of cell culture methods and enzymatic separation of epidermis and dermis [17, 18]. These bioengineered substitutes are composed of extracellular matrix (ECM) components and primary skin cells (keratinocytes, fibroblasts, and stem cells). These skin equivalents offer simulated-models for investigation of cell-cell and cell-ECM interactions, angiogenesis, skin irritation studies and so forth. In general, there are two major applications for HSE: i) for skin reconstruction and ii) as drug permeability and toxicity screening models [19]. Depending on the application, HSE can be designed as epidermis, dermis or composite full thickness skin models [20].

Commercial Epidermis Models

Commercial epidermal substitutes come in different forms of autologous/ allogenic keratinocytes such as cell-sheets or in conjugation with delivery systems like silicone or hyaluronic acid membrane.

Epicel (Genzyme, MA, USA) is one of the oldest cell delivery methods into the wound and is based on the use of autologous epidermal cells. Isolated keratinocytes from skin biopsy are cultured on top of fibroblast cells which act as a feeder layer and after sufficient confluency the cell sheets are transferred onto the wound. Epicel provides a permanent skin replacement for patients with extensive and full thickness burns involving greater or equal to 30% of total body surface area. Disadvantages of this method are long culturing time, fragility of cell sheets due to the lack of dermal layer and difficult handling [21, 22]. Epidex (Modex Therapeutic, Lausanne, Switzerland) is an autologous keratinocyte sheet from scalp hair follicles and is used for the treatment of chronic wounds. A retrospective study showed that Epidex treatment can lead to healing of 74% of the patients with chronic leg ulcers [23]. However, to reduce cell culture period, some delivery systems such as Myskin (Altrika Ltd., Sheffield, UK) have been developed to transfer pre-confluent autologous keratinocytes to the site of injury. Laserskin (Fidia Advanced Biopolymers, Abano Terme, Italy) was designed to transfer autologous keratinocyte cells cultured on a laser-microperforated hyaluronic acid membrane which allows cell migration from the membrane to the wound bed [8].

Commercial Dermis Models

As mentioned above, full-thickness skin equivalent is more promising for clinical applications, thereby, significant numbers of studies have investigated different approaches to design and produce reliable dermal skin equivalent.

Dermagraft (Organogenesis Inc., Canton, MA, USA) is a dermal analog which consists of neonatal fibroblasts cultured on an absorbable PLGA scaffold. As the scaffold degrades, the fibroblasts proliferate and synthesize growth factors and extracellular matrix. This product has achieved some successes on the treatment of diabetic foot ulcer [24, 25]. Other substitutes in this category are ICX-SKN (Intercytex, Manchester, UK) composed of a fibrin matrix seeded with neonatal human fibroblasts and Hyalograft-3D (Fidia Advanced Biopolymers, Italy) composed of esterified hyaluronic acid fibers seeded with autologous fibroblasts and covered by a silicone membrane which acts as an epidermis. Hyalograft-3D is mainly used in articular cartilage engineering; meanwhile, application in diabetic ulcer therapy has also been reported [25, 26].

Commercial Full-Thickness Skin Equivalent (Composite)

Dermo-epidermal or composite skin substitutes aim to simulate the skin histology and are the most advanced products to date. Both keratinocytes and fibroblasts are used to prepare bilayer living skin equivalents. The epidermal layer promotes regaining the protective barrier function and dermal layer enables rapid revascularization and physiological repair [27, 28].

Apligraft (Organogenesis Inc., Canton, MA, USA) is an early product of full-thickness skin equivalent composed of a dermal layer of bovine type I collagen and viable allogeneic neonatal fibroblasts, which produce additional matrix proteins, as well as viable allogeneic keratinocytes seeded on top of this dermal layer. It has been shown that Apligraft significantly improves the process of wound healing in chronic venous ulcers and chronic diabetic foot ulcers [29, 30]. Although this product does not trigger immune reactions, allogenic cells cannot survive after 1-2 months and the construct ultimately get rejected. Some other disadvantages include fragility of the construct, low shelf-life and the risk of human disease transfer [8].

Another composite substitute is OrCel (Ortec International, New York, USA). It is a bilayer cellular matrix comprising a top layer of non-porous bovine collagen laid over a porous layer of cross-linked type I bovine collagen. Similar to Apligraft, allogenic keratinocytes and fibroblasts from the same donor are cultured on and within the construct respectively. OrCell is an absorbable biocompatible matrix which provides a suitable environment for host cell infiltration. It has been reported that it causes reduced scarring and improves healing time when compared with Biobrane therapy [31].

The next step in designing permanent substitutes with the least immunological complication is to utilize autologous keratinocytes and fibroblast where it is possible. PermaDerm (Cincinnati Shriner's Hospital, Cincinnati, OH, USA) and Hyalograft 3D are such products. To apply PermaDerm, autologous cells are incorporated in collagen-GAG substrates. The cells are propagated in culture so as to lessen the required quantity of donor skin from the patient's own body. This product was successfully used in treatment of patients with full-thickness burns. The benefit of PermaDerm over split-thickness graft is the presence of dermal layer and the reduced need of donor skin [32]. To improve the appearance of grafted site, melanocytes can also be added to the keratinocyte culture to avoid hypopigmentation (*e.g.* ReCell) [33]. The Tissue-Tech Autograft System applies two different products Hyalograft 3D and Laser-skin as dermal and epidermal alternatives respectively. Autologous cells are grown on hyaluronic acid matrix. This substitute was proven to be effective in diabetic ulcer treatment [34].

Limitations of Commercially Available Skin Substitutes

The commercially available skin substitutes have several limitations. The first and the most common challenge is insufficient vascularization which results in initial nutritional crisis and consequently impaired regeneration. Several strategies have been used to improve the vascularization. A promising approach is co-seeding of the skin substitutes by endothelial progenitor cells or adipose stromal cells [35]. Recently, Klar and colleagues have introduced a prevascularized skin substitutes using stromal vascular fraction (SVF) derived from adipose tissue. This study showed the *de novo* formation of SVF-based microvascular networks and its integration into dermo-epidermal skin substitutes. Therefore, incorporation of adipose-derived SVF into an appropriate 3D construct can be a promising approach to allow prevascularization of the skin substitutes [36]. Another important limitation is the scar formation after grafting which results in different sorts of mechanical and aesthetic problems. In addition, the transmission of infection through the manufacturing processes can be another issue. Bacterial and fungal contamination could occur during cell culture despite meticulous control [7]. One other important problem restricting the practicality of skin substitutes is the lack of skin appendages like adipose tissue (in case of full thickness burns), hair follicles and sweat glands [37]. In addition, melanocytes and Langerhans cells are absent. Thereby, these substitutes provide inadequate pigmentation and immune responses. In order to overcome these restrictions, numbers of approaches are being applied with more emphasis on the use of stem cells as well as immune cells to develop a reliable model for skin disease studies.

SKIN STEM CELLS

Skin is the largest reservoir of various adult stem cells including mesenchymal stem cells (MSC), melanocyte stem cells (MelSC), hair follicle stem cells (HFSC) and adipose stem cells (ASC) [38]. These cells reside in different niches in the skin which include basal layer of epidermis, hair follicle, the base of the sebaceous glands and hypodermis (Table **2**). During wound healing process, stem cells play an important role in tissue reconstruction and it has been proven by some studies that epithelial stem cells in the hair follicle migrate to the epidermis and contribute to wound healing process [39, 40].

Table 2. Skin Stem Cells.

Skin Stem Cells	Location	Differentiated Progeny	Therapeutic Application
Interfollicular Epidermal Stem Cells	Epidermal basal layer	Suprabasal cells (*e.g.* Keratinocytes, Stratum corneum)	wound healing (long term maintenance)
Mesenchymal Stem Cells	Dermis, Bone-marrow, Amniotic fluid, Adipose tissue	Keratinocyte, Endothelial cells,	Cutaneous wound, Skin repair, Sweet gland regeneration Diabetic foot ulcer Critical limb ischemia
Hair Follicle Stem Cells	Bulb region & Hair germ at base of hair follicle	Keratinocytes, Melanocytes, Neurons, Glial cells, Smooth muscle cells	Skin regeneration
Melanocyte Stem Cells	Hair follicle bulge region & the Hair germ	Melanocytes	Skin pigmentation & Protection
Adipose-derived Stem Cells	Hypodermis, Adipose tissue in the body	Endothelial cells, Keratinocytes, Osteogenic differentiation potential	Wound healing, Neo-angiogenesis, Diabetic foot, venous ulcer, pressure ulcer

Mesenchymal Stem Cells

MSCs are multipotent stem cells and can be isolated from bone marrow, adipose tissue, amniotic fluid and dermis. Skin derived MSCs are primarily found in the dermal papilla and demonstrate site specific differentiation by adapting their functions to different environmental cues [41]. They contribute in wound healing process by secreting diverse growth factors including IGF-1, EGF, KGF, and mitogens to enhance fibroblast proliferation, ECM deposition as well as stem cell differentiation. For instance, through transdifferentiation into multiple cells like keratinocytes and endothelial cells, they participate in skin reepithelialization. In addition, MSCs can involve in dermis regeneration by affecting hair follicle morphogenesis and also with induction of inflammatory responses caused by macrophages [42, 43]. In this regard, MSCs are the most commonly used cells in preclinical and clinical studies. There are several different strategies to deliver MSCs to the wound including injection, spraying or through scaffold. The local injection of bone-marrow MSCs into an incisional full-thickness skin wound has been successfully demonstrated [44]. Nevertheless, the high rate of cell death after injection makes the scaffold a better candidate for cell delivery. Jeremias and colleagues reported the effectiveness of skin-derived MSCs in combination with collagen-based dermal substitutes (Integra and Pelnac) for the treatment of full thickness wound [45]. In a recent study it has also been shown that the scaffold loaded with stromal cell-derived factor (SDF)-1α, which is a stem cell homing factor, promotes MSCs recruitment to the injured site [46]. Together, scaffold offers an efficient environment for stem cell differentiation and delivery.

Hair Follicle Stem Cells

HFSCs and their progenitors directly participate in hair and skin regeneration. There are two main subpopulations of stem cells in hair follicles: quiescent group located in the bulb region and a subset located within the hair germ just below the bulge region [38]. Bulge stem cells have extensive differentiation potential with the ability to give rise into keratinocytes, neurons, melanocytes, and mesenchymal cells [47, 48]. Following full-thickness wound, HFSCs migrate to the site of injury and contribute in skin repair and re-epithelialization [49]. Successful application of HFSC in skin tissue engineering has been demonstrated when dissociated hair follicle cells and dermal cells were cultured in collagen gels and upon transplantation these cell mixture grown into a functional hair follicle [50]. This could open new avenues for including skin appendages to the engineered skin substitutes.

Adipose Stem Cells

Adipose is an interesting source of stem cells (ASC) and it has been shown that these cells can improve neo-epidermis formation by accelerating fibroblast proliferation and subsequently angiogenesis [51]. Combination of ASCs and fibroblasts in matrix has also been successfully used to accelerate wound healing of full-thickness cutaneous wounds [52]. The ability of these cells to grow on different matrices and their potential to differentiate into endothelial and epithelial cells makes them an excellent candidate for skin tissue engineering [53]. Polyvinyl alcohol (PVA) and gelatin nanofibers have been developed to direct ASCs differentiation into keratinocyte cells [54]. Together, preclinical studies have demonstrated that natural biomaterial scaffold such as collagen and cellulose derivatives seeded with ASCs present high potential for tissue repair by enhancing the epithelialization rate and downregulating the inflammatory responses [55, 56].

Induced Pluripotent Stem Cells

Induced pluripotent stem cells (iPSCs) with the ability to give rise to keratinocyte and melanocyte can be considered as a potential candidate for skin regeneration [57, 58]. The success in generating folliculogenic human epithelial stem cells from human iPSCs, which are able to reconstruct all hair follicle lineages, has been shown [59]. Besides, it has been reported that induced pluripotent stem cells-derived MSCs release exosomes that improve collagen synthesis and angiogenesis and consequently promote cutaneous wound healing [60]. However, despite the exciting progress has been achieved, the clinical application of iPSCs due to the risk of tumor formation through the use of retroviral vectors, inefficient cell re-programming and immunogenicity is still pending [61]. Nevertheless, iPSCs technology surmounts the ethical issues associated with embryonic stem cells and donor cells availability, therefore holds great promise for future clinical application in the field of wound repair.

RESEARCH ADVANCES OF SKIN TISSUE ENGINEERING

The majority of current toxicity studies is based on animal models and suffers from several limitations including their questionable biological relevance to human. Hence, numerous strategies have been investigated to design a credible model to increase its homology to native skin. The ongoing researches of skin equivalents have several applications in drug toxicity, permeability, chemical irritation and immunological studies.

Immunocompetent Model of Skin

Skin as our first line of defense is equipped with immune cells. In epidermis, specialized immune cells include αβ T cells, γδ T cells (minor subset in human) and Langerhans cells. In the dermis, macrophages, various dendritic cell (DC) subsets, innate lymphoid cells (ILCs), γδ T cells and αβ T cells are immunologically relevant cell types. Skin cells (including stem cells) and immune cells can influence each other's functionality and behavior during injuries and wound healing. This relationship provides protection to the body and present new strategies in vaccine research [62, 63].

Several research centers have developed immunocompetent skin equivalent models to study inflammatory skin diseases such as psoriasis, atopic dermatitis and skin sensitization *in vitro*. It has been shown that the different cell types such as DCs [64], T cells [65] as well as lymphatic capillaries [66] can be incorporated into tissue-engineered skin substitutes to partially simulate the construction of native skin. In a recent study, 3D skin equivalent was successfully populated with activated CD4+ T cells, Th1 and Th17 to develop a model of cross-talk between keratinocytes and T cells. This sort of communication is an essential factor in production of inflammatory phenotype which plays an important role in disease pathogenesis. In addition, in this model the application of drugs which target T-cells – keratinocytes cross-talk reduced the inflammatory phenotype of the cells. Therefore, this 3D skin construct can serve as a drug screening model for related skin disease [65].

Another immunocompetent model has been designed to investigate the mechanisms of skin sensitization (Fig. **2**). This model consists of a central layer of immune cells like DCs incorporated in agarose-fibronectin gel which is placed between an upper layer of keratinocytes and a bottom layer of fibroblasts. The application of topical skin sensitizer dinitrochlorobenzene (DNCB) has been shown to increase the activation of dendritic cells in this 3D co-culture model compared to a single culture of DCs (in agarose–fibronectin gel) and in the absence of keratinocyte and fibroblast layers. In 2D and 3D models none of the pro-inflammatory cytokines were upregulated in response to DNCB or Sodium dodecyl sulfate (SDS) stimulation which is likely due to keratinocytes differentiation status [64]. Since cross-talk between immune cells and dermo-epidermal cells is a critical step in inflammatory responses and wound healing process, immunocompetent models present potential application for drug screening, wound healing and skin disease studies [65, 67].

Skin Bioprinting Skills

Skin bioprinting is an innovative approach to develop a standardized skin model for evaluation of skin disease pathophysiology as well as an alternative substitute for wound treatment. It enables the simultaneous deposition of multiple cell types and ECM components with desired spatial configuration. Ink-jet printers are most commonly used in constructing tissues, but with this method, the damage of cells lead by shear stress has to be concerned. Therefore, the viscosity of biomaterials (ink) and cell density have to be strictly controlled. Two novel bioprinting approaches used for skin fabrication are microvalve- and laser-based printing methods [68, 69]. Koch employed laser-assisted bioprinting (LaBP) to construct multicellular grafts with NIH-3T3 fibroblasts and human immortalized keratinocytes (HaCaT) and with collagen as the cell container. The experimental setup consists two coplanar glass slides: the upper one with a laser absorbing material underneath (60nm thin gold layer); and the lower one containing cells and material to be transferred. In this study, a sheet of Matriderm (Dr. Suwelack Skin & Health Care, Billerbeck, Germany) was located on the lower glass, so when the laser pulses focused on the absorbing layer, the yielded pressure propelled the subjacent cells into the Matriderm layer. The printed cells showed high viability and proliferation ability, and formed gap junctions [70]. In a study by Lee, the composite 3D skin tissue has been bioprinted using valve-based method which enables to control the density and precise location of biomaterials (*e.g.* collagen) and cells (keratinocytes and fibroblasts). It has been shown that such methods facilitate the production of multilayer skin with the possibility to incorporate skin appendages and melanocyte [68, 69]. In addition, compared to traditional methods, bioprinting offers higher cell viability and other advantages like high throughput and the feasibility of *in situ*-printing [71, 72]. Hence, with further optimization of this technique we are able to develop more complex skin substitutes which can be used for wound treatment, as a model to enhance our understanding of the skin as an organ as well as to reduce our reliance on animals.

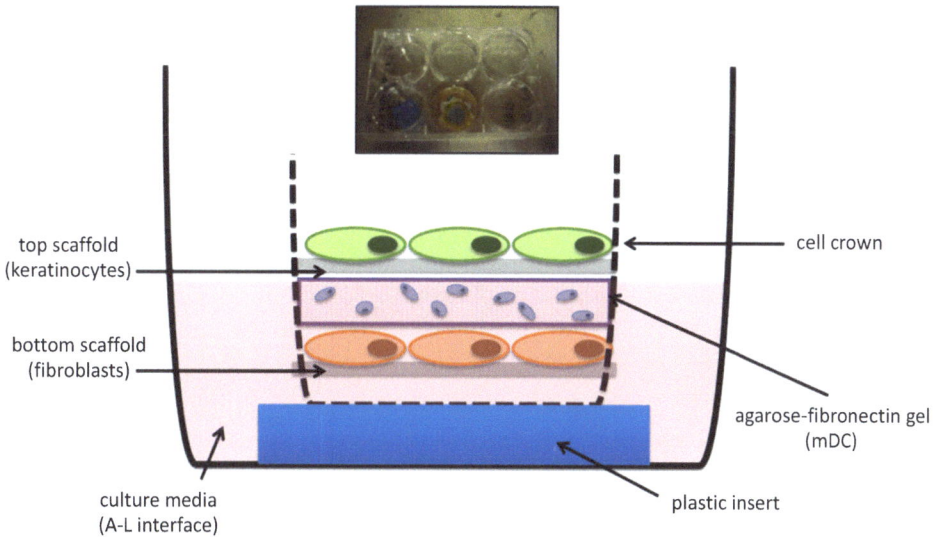

Fig. (2). Schematic of the complete 3D co-culture model comprising keratinocytes, fibroblasts and mDCs pre-seeded onto the appropriate matrix before being assembled together using CellCrown™ inserts—ensuring that the agarose gel was sandwiched between the upper (keratinocyte) and bottom (fibroblast) layers. This entire construct was then placed onto a plastic platform insert (blue square) with sufficient medium added to each well to ensure an air-liquid interface was maintained throughout the culture period [64].

Application of Electrospinning Skills in Skin Tissue Engineering

Electrospinning (ES) is an electrostatic fiber fabrication technique which generates nanoscale polymer fibers by the application of strong electric field on charged polymer solutions or melts. ES technique increases surface to volume ratio, not only by reduction of the size, but also by producing small pores in the materials. The porosity of the electrospun fibers is adjustable. In addition, electrospun fibers could also be designed for drug delivery, *e.g.* growth factors and antibiotics, which makes it more beneficial for tissue regeneration. Over the years, more than 200 materials have been electrospun for several applications including tissue engineering, drug delivery and biosensors [73]. Recent advances in using Chinese herbal drug (baicalein, BAI) combined with silk fibroin protein (SFP) and polyvinylpyrrolidone (PVP) based electrospun mats has shown some success for infected wound healing application. In an experimental mouse model with skin wound created for 1.2 cm×1.2 cm on back and infected by Staphylococcus aureus, the SFP/PVP/BAI nonwoven mat increased production of collagen fibers, enhanced angiogenesis, but meanwhile, reduced neutrophils infiltration, nitrite formation, and inhibited growth of wound bacteria [74]. Another study applied electrospun PCL membranes on a guinea pig full thickness excision wound healing model. The material enabled cell adhesion, migration and proliferation *in vivo*, and completely healed the skin defects in 35 days [75]. In summary, silk electrospinning scaffold can serve as an effective microenvironment to protect cell viability, cell proliferation and wound closure while it reduces the risk of burn wound infections which imposes a considerable cost on healthcare system.

CONCLUDING REMARKS AND FUTURE PERSPECTIVES

Tissue engineering of skin is an exciting area which already has made significant clinical impacts. Nevertheless, generating a complex dermo-epidermal substitute with the minimal scarring is still challenging. Current studies are trying to improve the outcomes by either encapsulating or coating the scaffolds or matrices with cytokines or other biological molecules to control cell infiltration, proliferation and differentiation into the constructs [76, 77]. Electrospinning and 3D bioprinting technology offer the opportunity for scaffold standardization and help to improve cell-scaffold interactions. Furthermore, stem cells can be utilized to facilitate the neovascularization and epithelialization of the wound healing process. The availability of adult stem cells and induced pluripotent stem cells (iPSCs) from the patient provides opportunities for eventually generating these structures without the risk of immune rejection for transplantation.

ABBREVIATIONS

ASCs   adipose derived stem cells

DC   dendritic cells

ECM	extracellular matrix
ESCs	embryonic stem cells
EGF	epidermal growth factor
HFSC	hair follicle stem cells
HSE	human skin equivalent
iPSCs	induced pluripotent stem cells
IGF	insulin growth factor
ILCs	innate lymphoid cells
KGF	keratinocyte growth factor
MSC	mesenchymal Stem cells
PVA	polyvinyl alcohol
PVP	polyvinylpyrrolidone
PGA	polyglycolic acid
PCL	polycaprolactone
PET	polyethylene terephthalate
PLGA	polylactic-*co*-glycolic acid
SVF	stromal vascular fraction

CONFLICT OF INTEREST

The authors declare no conflict of interest, financial or otherwise.

ACKNOWLEDGEMENTS

Declared none.

REFERENCES

[1]   Metcalfe AD, Ferguson MW. Tissue engineering of replacement skin: the crossroads of biomaterials, wound healing, embryonic development, stem cells and regeneration. J R Soc Interface 2007; 4(14): 413-37.
        [http://dx.doi.org/10.1098/rsif.2006.0179] [PMID: 17251138]

[2]   Gurtner GC, Werner S, Barrandon Y, Longaker MT. Wound repair and regeneration. Nature 2008; 453(7193): 314-21.
        [http://dx.doi.org/10.1038/nature07039] [PMID: 18480812]

[3]   Harvey C. Wound healing. Orthop Nurs 2005; 24(2): 143-57.
        [http://dx.doi.org/10.1097/00006416-200503000-00012] [PMID: 15902014]

[4]   Sen CK, Gordillo GM, Roy S, *et al.* Human skin wounds: a major and snowballing threat to public health and the economy. Wound Repair Regen 2009; 17(6): 763-71.
        [http://dx.doi.org/10.1111/j.1524-475X.2009.00543.x] [PMID: 19903300]

[5]   Clemens Van Blitterswijk. Tissue Engineering. Academic Press 2008.

[6]   Wong DJ, Chang HY. Skin tissue engineering. 2008.

[7]   Kamel RA, Ong JF, Eriksson E, Junker JP, Caterson EJ. Tissue engineering of skin. J Am Coll Surg 2013; 217(3): 533-55.
[http://dx.doi.org/10.1016/j.jamcollsurg.2013.03.027] [PMID: 23816384]

[8]   Shevchenko RV, James SL, James SE. A review of tissue-engineered skin bioconstructs available for skin reconstruction. J R Soc Interface 2010; 7(43): 229-58.
[http://dx.doi.org/10.1098/rsif.2009.0403] [PMID: 19864266]

[9]   Chen M, Przyborowski M, Berthiaume F. Stem cells for skin tissue engineering and wound healing. Crit Rev Biomed Eng 2009; 37(4-5): 399-421.
[http://dx.doi.org/10.1615/CritRevBiomedEng.v37.i4-5.50] [PMID: 20528733]

[10]   Zhong SP, Zhang YZ, Lim CT. Tissue scaffolds for skin wound healing and dermal reconstruction. Wiley Interdiscip Rev Nanomed Nanobiotechnol 2010; 2(5): 510-25.
[http://dx.doi.org/10.1002/wnan.100] [PMID: 20607703]

[11]   Melman L, Jenkins ED, Hamilton NA, *et al.* Early biocompatibility of crosslinked and non-crosslinked biologic meshes in a porcine model of ventral hernia repair. Hernia 2011; 15(2): 157-64.
[http://dx.doi.org/10.1007/s10029-010-0770-0] [PMID: 21222009]

[12]   Dai NT, Williamson MR, Khammo N, Adams EF, Coombes AG. Composite cell support membranes based on collagen and polycaprolactone for tissue engineering of skin. Biomaterials 2004; 25(18): 4263-71.
[http://dx.doi.org/10.1016/j.biomaterials.2003.11.022] [PMID: 15046916]

[13]   Supp DM, Boyce ST. Engineered skin substitutes: practices and potentials. Clin Dermatol 2005; 23(4): 403-12.
[http://dx.doi.org/10.1016/j.clindermatol.2004.07.023] [PMID: 16023936]

[14]   Shores JT, Gabriel A, Gupta S. Skin substitutes and alternatives: a review. Adv Skin Wound Care 2007; 20(9 Pt 1): 493-508.
[http://dx.doi.org/10.1097/01.ASW.0000288217.83128.f3] [PMID: 17762218]

[15]   Bello YM, Falabella AF, Eaglstein WH. Tissue-engineered skin. Current status in wound healing. Am J Clin Dermatol 2001; 2(5): 305-13.
[http://dx.doi.org/10.2165/00128071-200102050-00005] [PMID: 11721649]

[16]   Bano F, Barrington JW, Dyer R. Comparison between porcine dermal implant (Permacol) and silicone injection (Macroplastique) for urodynamic stress incontinence. Int Urogynecol J Pelvic Floor Dysfunct 2005; 16(2): 147-50.
[http://dx.doi.org/10.1007/s00192-004-1216-y] [PMID: 15378234]

[17]   Böttcher-Haberzeth S, Biedermann T, Reichmann E. Tissue engineering of skin. Burns 2010; 36(4): 450-60.
[http://dx.doi.org/10.1016/j.burns.2009.08.016] [PMID: 20022702]

[18]   Huang S, Fu X. Tissue-engineered skin: bottleneck or breakthrough. Int J Burns Trauma 2011; 1(1): 1-10.
[PMID: 22928152]

[19]   Zhang Z, Michniak-Kohn BB. Tissue engineered human skin equivalents. Pharmaceutics 2012; 4(1): 26-41.
[http://dx.doi.org/10.3390/pharmaceutics4010026] [PMID: 24300178]

[20]   Ehrenreich M, Ruszczak Z. Update on tissue-engineered biological dressings. Tissue Eng 2006; 12(9): 2407-24.
[http://dx.doi.org/10.1089/ten.2006.12.2407] [PMID: 16995775]

[21]   O'Conner NE, Mulliken JB, Banks-Schlegel S, Kehinde O, Green H. Grafting of burns with cultured epithelium prepared from autologous epidermal cells. Lancet 1981; 1(8211): 75-8.

[http://dx.doi.org/10.1016/S0140-6736(81)90006-4] [PMID: 6109123]

[22] Groeber F, Holeiter M, Hampel M, Hinderer S, Schenke-Layland K. Skin tissue engineering--*in vivo* and *in vitro* applications. Adv Drug Deliv Rev 2011; 63(4-5): 352-66.
[http://dx.doi.org/10.1016/j.addr.2011.01.005] [PMID: 21241756]

[23] Ortega-Zilic N, Hunziker T, Läuchli S, *et al.* EpiDex® Swiss field trial 2004-2008. Dermatology (Basel) 2010; 221(4): 365-72.
[http://dx.doi.org/10.1159/000321333] [PMID: 21071921]

[24] Marston WA, Hanft J, Norwood P, Pollak R. The efficacy and safety of Dermagraft in improving the healing of chronic diabetic foot ulcers: results of a prospective randomized trial. Diabetes Care 2003; 26(6): 1701-5.
[http://dx.doi.org/10.2337/diacare.26.6.1701] [PMID: 12766097]

[25] Stark HJ, Willhauck MJ, Mirancea N, *et al.* Authentic fibroblast matrix in dermal equivalents normalises epidermal histogenesis and dermoepidermal junction in organotypic co-culture. Eur J Cell Biol 2004; 83(11-12): 631-45.
[http://dx.doi.org/10.1078/0171-9335-00435] [PMID: 15679108]

[26] Flasza M, Kemp P, Shering D, *et al.* Development and manufacture of an investigational human living dermal equivalent (ICX-SKN). Regen Med 2007; 2(6): 903-18.
[http://dx.doi.org/10.2217/17460751.2.6.903] [PMID: 18034629]

[27] Bell E, Ehrlich HP, Buttle DJ, Nakatsuji T. Living tissue formed *in vitro* and accepted as skin-equivalent tissue of full thickness. Science 1981; 211(4486): 1052-4.
[http://dx.doi.org/10.1126/science.7008197] [PMID: 7008197]

[28] Rehim SA, Singhal M, Chung KC. Dermal skin substitutes for upper limb reconstruction: current status, indications, and contraindications. Hand Clin 2014; 30(2): 239-252, vii. [vii.].
[http://dx.doi.org/10.1016/j.hcl.2014.02.001] [PMID: 24731613]

[29] Falanga V, Sabolinski M. A bilayered living skin construct (APLIGRAF) accelerates complete closure of hard-to-heal venous ulcers. Wound Repair Regen 1999; 7(4): 201-7.
[http://dx.doi.org/10.1046/j.1524-475X.1999.00201.x] [PMID: 10781211]

[30] Veves A, Falanga V, Armstrong DG, Sabolinski ML. Graftskin, a human skin equivalent, is effective in the management of noninfected neuropathic diabetic foot ulcers: a prospective randomized multicenter clinical trial. Diabetes Care 2001; 24(2): 290-5.
[http://dx.doi.org/10.2337/diacare.24.2.290] [PMID: 11213881]

[31] Still J, Glat P, Silverstein P, Griswold J, Mozingo D. The use of a collagen sponge/living cell composite material to treat donor sites in burn patients. Burns 2003; 29(8): 837-41.
[http://dx.doi.org/10.1016/S0305-4179(03)00164-5] [PMID: 14636761]

[32] Boyce ST, Kagan RJ, Greenhalgh DG, *et al.* Cultured skin substitutes reduce requirements for harvesting of skin autograft for closure of excised, full-thickness burns. J Trauma 2006; 60(4): 821-9.
[PMID: 16612303]

[33] Gravante G, Di Fede MC, Araco A, *et al.* A randomized trial comparing ReCell system of epidermal cells delivery *versus* classic skin grafts for the treatment of deep partial thickness burns. Burns 2007; 33(8): 966-72.
[http://dx.doi.org/10.1016/j.burns.2007.04.011] [PMID: 17904748]

[34] Uccioli L. A clinical investigation on the characteristics and outcomes of treating chronic lower extremity wounds using the tissuetech autograft system. Int J Low Extrem Wounds 2003; 2(3): 140-51.
[http://dx.doi.org/10.1177/1534734603258480] [PMID: 15866838]

[35] Baranski JD, Chaturvedi RR, Stevens KR, *et al.* Geometric control of vascular networks to enhance engineered tissue integration and function. Proc Natl Acad Sci USA 2013; 110(19): 7586-91.
[http://dx.doi.org/10.1073/pnas.1217796110] [PMID: 23610423]

[36]   Klar AS, Güven S, Biedermann T, *et al.* Tissue-engineered dermo-epidermal skin grafts prevascularized with adipose-derived cells. Biomaterials 2014; 35(19): 5065-78.
[http://dx.doi.org/10.1016/j.biomaterials.2014.02.049] [PMID: 24680190]

[37]   Debels H, Hamdi M, Abberton K, Morrison W. Dermal matrices and bioengineered skin substitutes: a critical review of current options. Plast Reconstr Surg Glob Open 2015; 3(1): e284.
[http://dx.doi.org/10.1097/GOX.0000000000000219] [PMID: 25674365]

[38]   Hsu YC, Li L, Fuchs E. Emerging interactions between skin stem cells and their niches. Nat Med 2014; 20(8): 847-56.
[http://dx.doi.org/10.1038/nm.3643] [PMID: 25100530]

[39]   Taylor G, Lehrer MS, Jensen PJ, Sun TT, Lavker RM. Involvement of follicular stem cells in forming not only the follicle but also the epidermis. Cell 2000; 102(4): 451-61.
[http://dx.doi.org/10.1016/S0092-8674(00)00050-7] [PMID: 10966107]

[40]   Ito M, Liu Y, Yang Z, *et al.* Stem cells in the hair follicle bulge contribute to wound repair but not to homeostasis of the epidermis. Nat Med 2005; 11(12): 1351-4.
[http://dx.doi.org/10.1038/nm1328] [PMID: 16288281]

[41]   Jackson WM, Nesti LJ, Tuan RS. Concise review: clinical translation of wound healing therapies based on mesenchymal stem cells. Stem Cells Transl Med 2012; 1(1): 44-50.
[http://dx.doi.org/10.5966/sctm.2011-0024] [PMID: 23197639]

[42]   Ulivi V, Tasso R, Cancedda R, Descalzi F. Mesenchymal stem cell paracrine activity is modulated by platelet lysate: induction of an inflammatory response and secretion of factors maintaining macrophages in a proinflammatory phenotype. Stem Cells Dev 2014; 23(16): 1858-69.
[http://dx.doi.org/10.1089/scd.2013.0567] [PMID: 24720766]

[43]   Chen D, Hao H, Fu X, Han W. Insight into Reepithelialization: How Do Mesenchymal Stem Cells Perform? Stem Cells Int 2016; 2016: 6120173.

[44]   Chen L, Tredget EE, Wu PY, Wu Y. Paracrine factors of mesenchymal stem cells recruit macrophages and endothelial lineage cells and enhance wound healing. PLoS One 2008; 3(4): e1886.
[http://dx.doi.org/10.1371/journal.pone.0001886] [PMID: 18382669]

[45]   Jeremias TdaS, Machado RG, Visoni SB, Pereima MJ, Leonardi DF, Trentin AG. Dermal substitutes support the growth of human skin-derived mesenchymal stromal cells: potential tool for skin regeneration. PLoS One 2014; 9(2): e89542.
[http://dx.doi.org/10.1371/journal.pone.0089542] [PMID: 24586857]

[46]   Chen G, Tian F, Li C, *et al. In vivo* real-time visualization of mesenchymal stem cells tropism for cutaneous regeneration using NIR-II fluorescence imaging. Biomaterials 2015; 53: 265-73.
[http://dx.doi.org/10.1016/j.biomaterials.2015.02.090] [PMID: 25890725]

[47]   Yu H, Fang D, Kumar SM, *et al.* Isolation of a novel population of multipotent adult stem cells from human hair follicles. Am J Pathol 2006; 168(6): 1879-88.
[http://dx.doi.org/10.2353/ajpath.2006.051170] [PMID: 16723703]

[48]   Schneider M, Dieckmann C, Rabe K, Simon JC, Savkovic V. Differentiating the stem cell pool of human hair follicle outer root sheath into functional melanocytes. Methods Mol Biol 2014; 1210: 203-27.
[http://dx.doi.org/10.1007/978-1-4939-1435-7_16] [PMID: 25173171]

[49]   Levy V, Lindon C, Zheng Y, Harfe BD, Morgan BA. Epidermal stem cells arise from the hair follicle after wounding. FASEB J 2007; 21(7): 1358-66.
[http://dx.doi.org/10.1096/fj.06-6926com] [PMID: 17255473]

[50]   Toyoshima KE, Asakawa K, Ishibashi N, *et al.* Fully functional hair follicle regeneration through the rearrangement of stem cells and their niches. Nat Commun 2012; 3: 784.
[http://dx.doi.org/10.1038/ncomms1784] [PMID: 22510689]

[51] Matsuda K, Falkenberg KJ, Woods AA, Choi YS, Morrison WA, Dilley RJ. Adipose-derived stem cells promote angiogenesis and tissue formation for *in vivo* tissue engineering. Tissue Eng Part A 2013; 19(11-12): 1327-35.
[http://dx.doi.org/10.1089/ten.tea.2012.0391] [PMID: 23394225]

[52] Lu W, Yu J, Zhang Y, *et al.* Mixture of fibroblasts and adipose tissue-derived stem cells can improve epidermal morphogenesis of tissue-engineered skin. Cells Tissues Organs (Print) 2012; 195(3): 197-206.
[http://dx.doi.org/10.1159/000324921] [PMID: 21494022]

[53] Altman AM, Yan Y, Matthias N, *et al.* IFATS collection: Human adipose-derived stem cells seeded on a silk fibroin-chitosan scaffold enhance wound repair in a murine soft tissue injury model. Stem Cells 2009; 27(1): 250-8.
[http://dx.doi.org/10.1634/stemcells.2008-0178] [PMID: 18818439]

[54] Ravichandran R, Venugopal JR, Sundarrajan S, Mukherjee S, Forsythe J, Ramakrishna S. Click chemistry approach for fabricating PVA/gelatin nanofibers for the differentiation of ADSCs to keratinocytes. J Mater Sci Mater Med 2013; 24(12): 2863-71.
[http://dx.doi.org/10.1007/s10856-013-5031-1] [PMID: 23999881]

[55] Hassan WU, Greiser U, Wang W. Role of adipose-derived stem cells in wound healing. Wound Repair Regen 2014; 22(3): 313-25.
[http://dx.doi.org/10.1111/wrr.12173] [PMID: 24844331]

[56] Rodrigues C, de Assis AM, Moura DJ, *et al.* New therapy of skin repair combining adipose-derived mesenchymal stem cells with sodium carboxymethylcellulose scaffold in a pre-clinical rat model. PLoS One 2014; 9(5): e96241.
[http://dx.doi.org/10.1371/journal.pone.0096241] [PMID: 24788779]

[57] Aguiar C, Therrien J, Lemire P, Segura M, Smith LC, Theoret CL. Differentiation of equine induced pluripotent stem cells into a keratinocyte lineage. Equine Vet J 2016; 48(3): 338-45.
[http://dx.doi.org/10.1111/evj.12438] [PMID: 25781637]

[58] Ohta S, Imaizumi Y, Akamatsu W, Okano H, Kawakami Y. Generation of human melanocytes from induced pluripotent stem cells. Methods Mol Biol 2013; 989: 193-215.
[http://dx.doi.org/10.1007/978-1-62703-330-5_16] [PMID: 23483397]

[59] Yang R, Zheng Y, Burrows M, *et al.* Generation of folliculogenic human epithelial stem cells from induced pluripotent stem cells. Nat Commun 2014; 5: 3071.
[PMID: 24468981]

[60] Zhang J, Guan J, Niu X, *et al.* Exosomes released from human induced pluripotent stem cells-derived MSCs facilitate cutaneous wound healing by promoting collagen synthesis and angiogenesis. J Transl Med 2015; 13: 49.
[http://dx.doi.org/10.1186/s12967-015-0417-0] [PMID: 25638205]

[61] Okano H, Nakamura M, Yoshida K, *et al.* Steps toward safe cell therapy using induced pluripotent stem cells. Circ Res 2013; 112(3): 523-33.
[http://dx.doi.org/10.1161/CIRCRESAHA.111.256149] [PMID: 23371901]

[62] Gay D, Kwon O, Zhang Z, *et al.* Fgf9 from dermal γδ T cells induces hair follicle neogenesis after wounding. Nat Med 2013; 19(7): 916-23.
[http://dx.doi.org/10.1038/nm.3181] [PMID: 23727932]

[63] Heath WR, Carbone FR. The skin-resident and migratory immune system in steady state and memory: innate lymphocytes, dendritic cells and T cells. Nat Immunol 2013; 14(10): 978-85.
[http://dx.doi.org/10.1038/ni.2680] [PMID: 24048119]

[64] Chau DY, Johnson C, MacNeil S, Haycock JW, Ghaemmaghami AM. The development of a 3D immunocompetent model of human skin. Biofabrication 2013; 5(3): 035011.
[http://dx.doi.org/10.1088/1758-5082/5/3/035011] [PMID: 23880658]

[65]   van den Bogaard EH, Tjabringa GS, Joosten I, *et al.* Crosstalk between keratinocytes and T cells in a
        3D microenvironment: a model to study inflammatory skin diseases. J Invest Dermatol 2014; 134(3):
        719-27.
        [http://dx.doi.org/10.1038/jid.2013.417] [PMID: 24121402]

[66]   Marino D, Luginbühl J, Scola S, Meuli M, Reichmann E. Bioengineering dermo-epidermal skin grafts
        with blood and lymphatic capillaries. Sci Transl Med 2014; 6(221): 221ra14.
        [http://dx.doi.org/10.1126/scitranslmed.3006894] [PMID: 24477001]

[67]   Peters JH, Tjabringa GS, Fasse E, *et al.* Co-culture of healthy human keratinocytes and T-cells
        promotes keratinocyte chemokine production and RORγt-positive IL-17 producing T-cell populations.
        J Dermatol Sci 2013; 69(1): 44-53.
        [http://dx.doi.org/10.1016/j.jdermsci.2012.10.004] [PMID: 23127421]

[68]   Michael S, Sorg H, Peck CT, *et al.* Tissue engineered skin substitutes created by laser-assisted
        bioprinting form skin-like structures in the dorsal skin fold chamber in mice. PLoS One 2013; 8(3):
        e57741.
        [http://dx.doi.org/10.1371/journal.pone.0057741] [PMID: 23469227]

[69]   Lee V, Singh G, Trasatti JP, *et al.* Design and fabrication of human skin by three-dimensional
        bioprinting. Tissue Eng Part C Methods 2014; 20(6): 473-84.
        [http://dx.doi.org/10.1089/ten.tec.2013.0335] [PMID: 24188635]

[70]   Koch L, Kuhn S, Sorg H, *et al.* Laser printing of skin cells and human stem cells. Tissue Eng Part C
        Methods 2010; 16(5): 847-54.
        [http://dx.doi.org/10.1089/ten.tec.2009.0397] [PMID: 19883209]

[71]   Lee W, Debasitis JC, Lee VK, *et al.* Multi-layered culture of human skin fibroblasts and keratinocytes
        through three-dimensional freeform fabrication. Biomaterials 2009; 30(8): 1587-95.
        [http://dx.doi.org/10.1016/j.biomaterials.2008.12.009] [PMID: 19108884]

[72]   Binder KW, Zhao W, Aboushwareb T, Dice D, Atala A, Yoo JJ. *In situ* bioprinting of the skin for
        burns. J Am Coll Surg 2010; 211(3): S76.
        [http://dx.doi.org/10.1016/j.jamcollsurg.2010.06.198]

[73]   Bhardwaj N, Kundu SC. Electrospinning: a fascinating fiber fabrication technique. Biotechnol Adv
        2010; 28(3): 325-47.
        [http://dx.doi.org/10.1016/j.biotechadv.2010.01.004] [PMID: 20100560]

[74]   Chan WP, Huang KC, Bai MY. Silk fibroin protein-based nonwoven mats incorporating baicalein
        Chinese herbal extract: preparation, characterizations, and *in vivo* evaluation. J Biomed Mater Res B
        Appl Biomater 2017; 105(2): 420-30.
        [http://dx.doi.org/10.1002/jbm.b.33560] [PMID: 26540289]

[75]   Augustine R, Dominic EA, Reju I, Kaimal B, Kalarikkal N, Thomas S. Electrospun poly(ε-
        caprolactone)-based skin substitutes: In vivo evaluation of wound healing and the mechanism of cell
        proliferation. J Biomed Mater Res B Appl Biomater 2015; 103(7): 1445-54.
        [http://dx.doi.org/10.1002/jbm.b.33325] [PMID: 25418134]

[76]   Jin G, Prabhakaran MP, Ramakrishna S. Photosensitive and biomimetic core-shell nanofibrous
        scaffolds as wound dressing. Photochem Photobiol 2014; 90(3): 673-81.
        [http://dx.doi.org/10.1111/php.12238] [PMID: 24417712]

[77]   Morimoto N, Yoshimura K, Niimi M, *et al.* Novel collagen/gelatin scaffold with sustained release of
        basic fibroblast growth factor: clinical trial for chronic skin ulcers. Tissue Eng Part A 2013; 19(17-18):
        1931-40.
        [http://dx.doi.org/10.1089/ten.tea.2012.0634] [PMID: 23541061]

Frontiers in Biomaterials, 2017, *Vol. 5*, 131-169   131

Stem Cells and Scaffolds: Strategies for Musculoskeletal System Tissue Engineering

Mahboubeh Nabavinia[1,2,3,§], **Ding Weng**[4,5,7,§], **Yi-Nan Zhang**[6] and **Wanting Niu**[4,5,*]

[1] *Harvard-MIT Division of Health Sciences and Technology, Massachusetts Institute of Technology, Cambridge, USA*

[2] *Biomaterials Innovation Research Center, Department of Medicine, Brigham and Women's Hospital, Harvard Medical School, Cambridge, USA*

[3] *Chemical Engineering Department, Sahand University of Technology, Tabriz, Iran*

[4] *Tissue Engineering Labs, VA Boston Healthcare System, Boston, USA*

[5] *Department of Orthopedics, Brigham and Women's Hospital, Harvard Medical School, Boston, USA*

[6] *Institute of Biomaterials and Biomedical Engineering, University of Toronto, Toronto, Canada*

[7] *Department of Mechanical Engineering, State Key Lab of Tribology, Tsinghua University, Haidian Qu, Beijing Shi, China*

Abstract: Musculoskeletal system mainly includes bone, cartilage, tendon, ligament, and skeletal muscle, which supports the body and connects tissues and organs together. Musculoskeletal disorders, including degenerative diseases and trauma injuries lead to the loss of part tissue, and always accompanied with pain. With the development of biomaterials, developmental biology, nanomedicine and orthopedic therapy strategies, tissue engineering of musculoskeletal system has been well investigated in the past two decades. Many musculoskeletal tissue repair strategies have achieved success in animal models, with part of them entered clinical trials which will benefit patients in the future.

Keywords: Bone, Cartilage, Decellularized scaffolds, Hybrid materials, Ligament, Natural polymers, Skeletal muscle, Stem cells, Synthetic polymers, Tendon, Tissue engineering.

INTRODUCTION

The musculoskeletal system includes the bones of the skeleton and cartilages

---

[*] **Corresponding author Wanting Niu:** VA Boston Healthcare System; Brigham and Women's Hospital, Harvard Medical School, Boston, USA; Tel: +1-617-637-6609; E-mail: wniu@partners.org
[§] These authors contributed equally to this chapter

Mehdi Razavi (Ed.)
All rights reserved-© 2017 Bentham Science Publishers

(hyaline cartilage, elastic cartilage, and fibrocartilage), tendons, ligaments and muscles which attach to the bones and stabilize them. The musculoskeletal system supports the whole human body, and enables locomotion. Musculoskeletal disorders could manifest in senior people due to degenerative diseases, such as arthritis, osteoporosis, and muscle atrophy; Physical active population, including athletes and military service people are also take high risks of musculoskeletal injuries, *e.g.* tendonitis, rotator cuff tear and skeletal muscle injuries. Pain, ranges from mild to severe, is the most common syndrome of musculoskeletal tissue disorders. Severe pain could limit the patients' movements, and severe injuries can also result in dysfunction of the damaged tissue. Tissue engineering aims to reconstruct injured tissue with long-lasting restoration of function, with the tools of biomaterials scaffolds, growth factors and bioreactors, and cells. The objective of this chapter is to review recent research advances of engineered bone, cartilage, tendon and ligament and skeletal muscle based on stem cells.

BONE TISSUE ENGINEERING

Bone tissue engineering (BTE) provides the opportunities for bone repair, regeneration and disease treatment. Based on differentiate and undifferentiated status, the bone cell types can be divided into osteoblast progenitor cells and stem cells, respectively [1 - 6]. Stem cells enable to differentiate to the specific cell phenotypes of bone tissues, and interact with biocompatible scaffold to successfully regenerate to the damaged region in clinical therapies [1]. In addition, stem cells offer broad application due to the off-the-shelf availability, high harvesting efficiency, good expansion capacity, cryopreservation, and high osteogenic potential. The main stem cell types used for BTE are mesenchymal stem cells (MSCs), embryonic stem cells (ESCs), and induced pluripotent stem cells (iPSCs). In addition to the primary cell line, the physicochemical properties of the supporting porous scaffold biomaterials can influence stem cell differentiation pathways [7]. In this section, we highlight the significant developments of using stem cells in BTE and focus on the engineered biomaterials for maturation of bone engineered cellular constructs.

Engineered Bone Based on Mesenchymal Stem Cells

Nowadays, BMPs in combination with stem cells and/or scaffolds gain more and more attention for their use in BTE [8 - 11]. The description of MSCs as fibroblastoid cells was reported by Friedenstein *et al.* in 1966 [12], and Owen in 1988 [13]. Since then, the MSCs have been considered for tissue engineering applications. Fig. (**1**) shows the cell therapy registered in clinical trials. Up to 2014, more than 110 clinical trials have been registered to evaluate MSC prospections for disease treatment, which significantly surpass other types of cells, and the trend keeps positive. About half of these clinical applications reported MSCs for treating orthopedics diseases and defects [6]. MSCs can be isolated from postnatal tissue and virtually postnatal organs and tissues but main sources are taken from the bone marrow (BMSCs) [14 - 16], adipose tissue [17,18], dental pulp [19, 20], umbilical cord blood [21 - 23] with osteogenic potential. MSCs was firstly used in the clinical trial in 2001 for bone formation in large bone defects [24]. Many clinical trials have been reported that MSCs can be incorporated with various scaffolds [20]. These scaffolds, made by porous biocompatible and degradable materials, can provide additional support and 3D architecture for stem cell growth, differentiation and proliferation. The physicochemical and mechanical properties (strength, strain and stiffness) of porous scaffold can also influence the hMSCs' differentiation [25 - 27]. Murphy *et al.* reported that the mechanical stiffness and nanoscale topography of materials can affect the differentiated human mesenchymal stem cells (hMSCs) on their functions [7]. Li *et al.* reported that the differentiation of bone marrow and adipose tissue derived MSCs prefers the softer substrate compared to the stiffer one. Interestingly, stiffness has no effect on these MSCs' proliferation [27].

Of note, turbinate mesenchymal stem cells (TMSCs) have been isolated from inferior turbinate tissue with a minimum surgical procedure, and identified as a potential therapeutic source. The human TMSCs (hTMSCs) collected from removed turbinate tissue during surgery exhibited similar characterizations as BMSCs, but with higher proliferation rate and less erythrocytes contamination. Kwon *et al.* encapsulated hTMSCs in a methoxy polyethylene glycol–polycaprolactone block copolymer (MPEG–PCL) solution, and injected this cell laden *in situ*-forming hydrogel into subcutaneous dorsum of nude mice. The histology staining and gene expression test certified the *in vivo* mineralized bone-like tissue formation [28]. In another study, hTMSCs were incorporated in silk fibroin-gelatin (SF-G) 3-D printing bioink and induced *in situ* gelation by enzymatic crosslinking and physical crosslinking *via* mushroom tyrosinase and sonication respectively. Gene expression profiles assessed the osteogenic, chondrogenic and adipogenic abilities of hTMSCs in this SF-G matrix [29].

To re-construct large bone defects, the main challenge is to refill the defected area with porous biomaterials which permit angiogenesis and revascularization. Park *et al.* grafted a complex, sandwich-like filling structure with the employment of multi-head tissue/organ building 3-D printing system. Polycaprolactone (PCL) was applied as the framework material, while dental pulp derived stem cells (DPSCs) encapsulated alginate hydrogels and collagen I hydrogels were designed to fill the gaps between PCL fibrosis. Vascular endothelial growth factor (VEGF) and bone morphogenetic protein (BMP)-2 were incorporated in alginate gels and collagen I gels respectively. This structure allows burst release of VEGF from the central zone to induce vascularization of DPSCs, then slow release of BMP-2 which contributes for bone formation. The DPSC-scaffold complex was implanted subdermally in the backs of BALB/c-nu/nu immunodeficient mice for four weeks. The observation of red blood cells in the peripheral portion of the constructs confirmed the formation of vascular tissues, and enhanced bone regeneration was also found in those pre-vascularized regions [30]. Also with the purpose of better mimicking nature vascularized bone, Cui *et al.* stated an art on the platforms of 3-D bioprinting systems and perfusion bioreactor with polylactide (PLA) and gelatin methacrylate (GelMA) as the raw materials. With the FDM bioprinter, PLA was arrayed as the hard portion of the artificial bone with honeycomb structure. Thereafter, human MSC and human umbilical vein endothelial cells (HUVECs) laden GelMA was photocrosslinked among the space in PLA array on the SLA bioprinter to mimic the ECM part. In order to let the encapsulated cells get enough gas and mass transportation, and let HUVECs experience continuous shear stress during their development of vascular lumen, a perfusion bioreactor was also applied to provide a dynamic culture environment. Their results showed the encapsulated HUVECs formed annulus pattern around the channels and the hollow structure allows native vessel invasion and integration. The osteogenic markers and calcium deposition were also measured elevated as the benefits of using bioreactor [31].

The MSCs' osteogenic differentiation can be further promoted by additional biochemical reagent treatment on the scaffold. Cell growth or stimulating factor can be cooperated in the scaffold to orchestrate cell fate, *e.g.* material mediated sequestering can harness BMP for BTE [9, 10, 32, 33]. Advantages and disadvantages of three main groups of biomaterials (natural biopolymers, synthetic biopolymer, and ceramics) were summarized in Table **1**.

Table 1. Advantages and disadvantages of main groups of biomaterials.

Biomaterials	Descriptions	Advantages	Disadvantages
Natural polymers	Mainly include proteins (collagen, gelatin, elastin, laminin, fibroin …) polysaccharides (chitosan, alginate, hyaluronic acid, pectin…) and starch based (bacterial cellulose, dextran…)	High biocompatibility and biodegradability Great interaction with cells	Variable properties after processing Temperature sensitive (especially protein) Difficulty for sterilization Showing inflammatory *in vivo* Mechanical weakness and instability Expensive

(Table 1) contd.....

Biomaterials	Descriptions	Advantages	Disadvantages
Synthetic polymers	Main groups contain: polyesters (polyglycolic acid, polylactic acid, polycaprolactone...) Bioresorbable polymer (urethane), Polypropylene fumarate	Availability and low cost FDA approved Can be processed and shaped easily to desired specifications with consistent quality	low cell attachment and interaction Possible interference with the healing degradation products can have negative effects on the healing environment
Ceramic	Most common materials are: tricalcium phosphate, hydroxyapatite, Bioglass...	High similarities with bone mineral part biocompatibility and osteoconduction	High fracture potential Particularly low cell attachment
Composites	Combination polymers (Naturally or synthetic) with ceramic	Adjusting chemical, physical and mechanical properties the target tissue Great cell adhesion and spread	Complexity of fabricating process

Another important issue which has been recently attended is the non-static properties at the cell/material interface. Due to the scaffold degradation, the microenvironment at the interface is dynamic and the degraded by-products can influence the cell fate. Mineral components, such as calcium, magnesium ions released from the surface of the inorganic materials be vary the stem cell function [34]. The by-products from the degraded polymer based materials, such as agarose, gelatin or poly(γ-glutamic acid), can alter the stem cell differentiation pathway [35].

Engineered Bone Based on Embryonic Stem Cells

Due to the high potential of rapid proliferation over long periods and capability of differentiation into most cell types, ESCs provide promising supply of stem cells and they have the high capability to the generation of osteogenic cells. Adding the proper supplements to culture medium, such as ascorbic acid, β-glycerophosphate, and dexamethasone, can promote ESCs to differentiate into osteoblast linage [36 - 38]. Manipulating the chemical properties and 3D architecture of the scaffolds have been used to promote ECSs' osteogenesis and formation of bone-like structure.

Chemically modified gelatin hydrogels which was cross-linked by glyceraldehyde is ideal for ESC osteogenic differentiation [39]. In addition, the static cultured or co-cultured cells on 2D or 3D were inhibited from proliferation and differentiation due to the limited stimulating factors and nutrient transportation as well as dynamic cell interaction. Hwang *et al.* reported that they generated a functional

3D bone-like tissue by encapsulated ESCs in alginate hydrogels and in a rotated culture bioreactor [40]. Marolt *et al.* also reported that human ESC-derived mesenchymal progenitors induce the formation of a uniform and stable bone-like tissue using a 3D osteoconductive scaffold with interstitial flow [41]. Another study testified the osteoconductive and potential osteoinductive effects of calcium phosphate coated hydroxyapatite mineral particles (MPs) on murine D3 ESC aggregates. The results revealed that in combination of soluble osteoinductive biomolecules, the MPs enhanced the osteochondrogenic differentiation and mineralization; but reduced the pluripotency of the ESCs so that reduced the risk of teratoma formation [42]. These results indicate the importance of mechanical stimulation and dynamic flow in porous scaffolds for bone tissue regeneration.

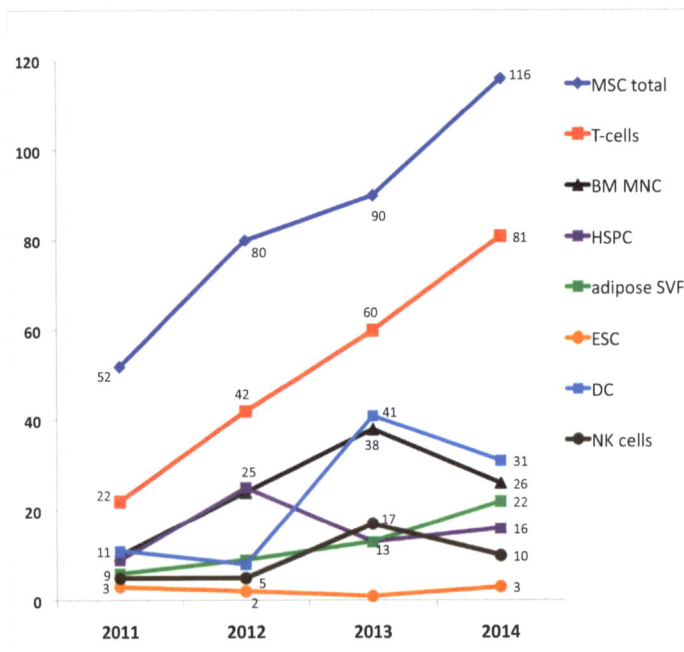

Fig. (1). Cell therapy registered in clinical trials [6].

Engineered Bone Based on Induced Pluripotent Stem Cells

Due to the ethical dilemmas and political restriction of ESCs, using induced pluripotent stem cells (iPSCs) become an alternative approach for cell and tissue regeneration [43, 44]. It is reported that the human iPSCs (hiPSCs) performed similar properties to the hESCs and can be served for bone tissue engineering [45, 46]. Mouse specific iPSCs were reported to assess MSCs and induce osteoblast differentiation *in vitro* [47]. In addition, material can promote the function of iPSCs as an extracellular matrix and porous scaffold. Wang compared human induced pluripotent stem cell-derived mesenchymal stem cells (hiPSC-MSCs)

derived from bone marrow (BM-hiPSC-MSCs) and from foreskin (FS-hiPS-
-MSCs), with dental pulp stem cells (hDPSCs) and bone marrow MSCs
(hBMSCs) for BTE in an injectable calcium phosphate cement (CPC) scaffold.
All four kinds of cells underwent osteogenic differentiation in hydrogel fibers, but
BM-hiPSC-MSCs, hDPSCs and hBMSCs showed more promising data with bone
formation related gene expression (alkaline phosphatase, runt-related transcription
factor, collagen I, and osteocalcin) [48]. In another study, hiPSCs were seeded in
decellularized bovine bone substitutes and enabled to differentiate into
mesenchymal progenitor cells in a perfused bioreactor, which indicates strong
potential for bone generation [49, 50]. Jin *et al.* also reported that hiPSCs can
induce *in vivo* bone formation in a 3D poly(caprolactone) (PCL) porous scaffold.
Their histological results indicated the bone ECM production and mineral
deposition when the cell laden scaffolds had been implanted subcutaneously into
athymic mice for 4 weeks [51]. However, because of the genomic instability, risk
of tumor formation, immune response and variability among iPSC clones and
differentiation capacity, many trials have failed to translate iPSCs for clinical
practice. Meanwhile, there are many new methods are under investigation to
overcome the risks, such as reprogramming with small DNA-modifying
molecules and using suicide genes to remove undifferentiated iPSCs post
implantation [52].

CARTILAGE TISSUE ENGINEERING

As cartilage tissue's avascular nature, it has very limited capacity for spontaneous
healing. Cartilage tissue engineering (CTE) was proposed for the first time by
Green in 1977 [53]. It can be defined to engineer the cartilage tissue using 3D
porous scaffold with proper structure and function [54]. Regarding to cell type, an
autologous chondrocyte implantation (ACI) is gold standard due to the risks
associated with allogenic strategies, such as triggering an immunogenic response
or transferring diseases. However, the important stages of ACI employed CTE
includes: 1) extract arthroscopically from the patient's healthy articular cartilage;
2) expand the extracted cells *in vitro* without phenotype changes (chondrocytes
are expanded *in vitro* for approximately four to six weeks); 3) second surgery to
implant a sufficient number and amplified chondrocytes to the damaged area of
the patient. The procedures are invasive and time consuming, which is also
controlled by the health situation and age of the patient. Achieving proper new
articular cartilage depends on the proper growth of transplanted cells in their new
microenvironment [55]. In order to overcome these challenges, many
investigators have recently focused on using stem cells that can differentiate to
chondrocyte [9]. Many types of stem cells are used for CTE and their availability,
proliferation rate, and differentiation efficiency are the most important factors.
Among various types of stem cells, MSCs, ESCs, and iPSCs are commonly

considered for CTE application. With the support of natural and synthetic porous scaffold for stem cell derived chondrocyte formation, some engineered scaffold based therapeutic approaches have been used in the clinical praxis. In this section, we will discuss the usage of these biomaterials to induce the stem cell differentiation to chondrocytes.

Engineered Cartilage Based on Mesenchymal Stem Cells

Due to the properties of easy availability, inherent chondrogenic ability, and cell homing potential, MSCs are considered as a key candidate for CTE. Martin *et al.* reported that ~16% of all the cell-based cartilage repairs employed MSCs [56]. MSCs can be extracted from many tissues and they have different proliferation rate and differentiation capacity [11]. MSCs derived from adipose tissue proliferate about 500-fold faster than those from bone marrow [10, 40]. Regarding to practical advantages, MSCs derived from synovial membranes have great chondrogenic potential [54]. Local microenvironment, extracellular matrix, and growth factors play important roles in differentiating MSCs to different cartilage cells. TGF-β1 is the most important growth factor, which is a potent inducer of chondrogenic differentiation which favors to the formation of ectopic cartilage *in vivo*. TGF-β1 also promotes MSCs proliferation in certain biomaterials [57 - 59]. Biomaterials benefit MSC differentiation and proliferation by providing tissue simulating 3D environment, especially ligands for cell adhesion.

In general, the biomaterials which have been applied for cartilage scaffolds have viscoelastic behavior. Semicrystalline polymers with this behavior can mimic the mechanical property of the cartilage [60]. Among the most types of biomaterials, hydrogels could provide unique properties, such as injectable, similar stiffness/structure as nature ECM of some types of tissue, easy manipulation for surgery, and high percentage of water content [61]. The swollen structure of hydrogel provides similar conditions as natural cartilage ECM on the points of mass transfer and mechanical force cushion. Furthermore, these structures promote cell homogeneous distributions [62 - 64]. Some hydrogels have been employed for clinical applications, as summarized in Table **2** [65], and the gel formation mechanisms, including permanent covalent crosslinking and physical reversible crosslinks were exhibited in Fig. (**2**).

Table 2. Main types of hydrogel matrices used in CTE [63].

Type	Component	Commercial Product Name
Protein	Collagen	MACI®, Maix®, atelocollagen®, MaioRegen® FibriCol®
	Fibrin	Tissucol kit®
	Silk, Fibroin	

(Table 2) contd.....

Type	Component	Commercial Product Name
	Elastin Laminin	MAPTrixTM
Polysaccharides	Hyaluronic acid	HYAFF-11®
	Chitosan Dextran	BST-CarGel®
	Cellulose	
	Alginate/Agarose	Xizia Biotech®
Synthetic	Poly(Lactic-coglycolic acid) Polyethylene glycol acid	Bio-Seed®-C BioTissue®
	Polylactic acid	
	Polyethylene glycol PLGA/calciumsulfate/ PGA fibers	Regenexx® TruFit®

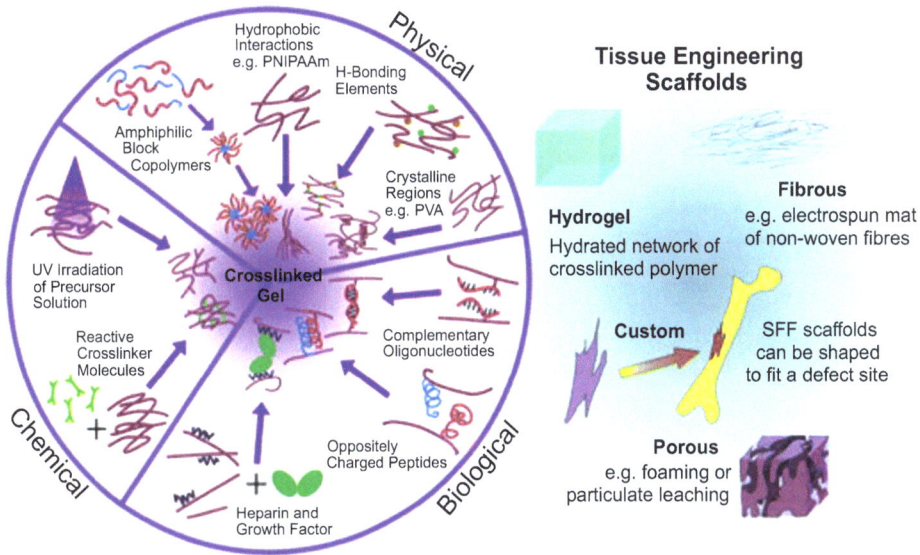

Fig. (2). Hydrogels in tissue engineering [66] (with permission from the Royal Society of Chemistry).

Engineered Cartilage Based on Embryonic Stem Cells and induced Pluripotent Stem Cells

The knowledge of ESC chondrogenic potential was learnt from teratomas, because the cartilage islands in a disorganized manner exist in this tumor tissue [67]. To date, several factors, including small molecules, growth factors (BMP 2 or 4, TGF-β1 or 3), microenvironment, and biophysical stimulations, have been considered critical for chondrogenic potential of ESCs [68 - 70]. Levenberg and colleagues showed how the factors of retinoic acid, transforming growth factor β, activin-A, or insulin-like growth factor influence hESC differentiation and organization [71].

In a study by Vats *et al.*, co-cultured hESCs and chondrocytes were seeded into PLLA scaffolds and implanted subcutaneously in the backs of immunodeficient mice. They investigated the effects of microenvironment on cell differentiation both *in vitro* and *in vivo*. After five weeks, implanted co-cultured hESCs developed into cartilaginous tissues that were positive for glycosaminoglycans (GAG) and collagen II. In addition, collagen I and calcification were also detected, suggesting the existence of fibrocartilage and growth plate cartilage [69]. Similarly, hESCs were induced into chondrogenically-committed cells by chondrocytes and encapsulated in poly(ethylene-glycol)-diacrylate (PEGDA) photocrosslinkable hydrogel which was modified with arginineglycine-aspartate (RGD) to form cell-hydrogel complex. The results of *in vitro* study found that within 3-week appropriate culture, neocartilage-like tissue formed with basophilic ECM deposition. After a certain period, histological staining revealed cartilage formation resulted from the cell encapsulated PEGDA gel which was implanted subcutaneously into athymic mice [72]. Toh *et al.* demonstrated that transplanted hESC-derived chondrogenic cells which were encapsulated in hyaluronic acid (HA)-based hydrogels, maintained long-term viability with no evidence of tumorigenicity. Those cells exerted a dual function: repairing cartilage tissue by themselves and stimulating paracrine *via* promoting the ingrowth of endogenous cells [73].

Established iPS lines which could be efficiently differentiated into ECM secreting chondrocytes, have been investigated as a promising cell source for cartilage reconstruction. Diekman and his colleagues considered creating tissue-engineered cartilage constructs with pre-chondrogenic differentiated and purified iPSCs derived from mouse tail fibroblast *in vitro* with agarose gel. They induced chondrogenic differentiation by treating micromass cultures with BMP-4 and then sorted differentiated cells based on expressing green fluorescent protein (GFP) by flow cytometry under control of type II collagen (Col2) promoter/enhancer [74]. With using CRISPR/Cas9 technique, Brunger *et al.* deleted interleukin 1 (IL-1) receptor 1 (Il1r1) gene in murine iPSCs because IL-1 elevates in injured tissue which prevents stem cell differentiation. On the pellet chondrogenic differentiation model, the gene edited iPSCs showed resistant to the IL-1 induced tissue degradation [75]. It has been tested on murine iPSCs that knockdown of cell cycle inhibitor p21 could also enhance cartilage-specific Col 2 expression, thus improve cartilage formation [76].

TENDON AND LIGAMENT TISSUE ENGINEERING

Tendon and ligament are both dense connective tissues but tendon offers muscle to bone, while ligament provides bone to bone connection. Tendon and ligament injuries happen frequently during sports and incidents. Their natural healing is quite slow and always results in scar tissue formation which further leads to mechanical instability and re-injury in the future. Some injuries involve tendon/ligament loss or gap, especially in the extensor tissues which are the most challenging cases to reconstruct with surgeries [77].

The overall outcomes of surgical treatments have not been dramatically improved in the past 50 years although many innovations have been developed [78]. In many cases, surgical repairs cannot help in fully restoring biological functions of injured tendon and ligaments due to the fibrous adhesions at tendon-tendon or tendon-bone conjunctions. Therefore, regeneration techniques are necessary to develop more effective connective tissue including tendon and ligament. Tissue engineering aims to reconstruct the injured tendon/ligament with biofunctional extracellular matrix to provide a framework for tissue ingrowth and integration.

Achilles tendon and anterior cruciate ligament (ACL) are always be selected as model tissues in tendon and ligament tissue engineering. The patients with Achilles tendon injuries feel strong pain and are not possible to stand and walk on tiptoe after the Achilles tendon ruptures. The surgical methods to treat Achilles tendon ruptures including open and percutaneous (closed) methods [79]. Anterior cruciate ligament plays critical roles as keeping the knee joint stable and enabling movements, but it is the most frequently damaged ligament due to vigorous activities. In the U.S.A. alone, the ACL epidemiologic studies estimated that the yearly incidence ACL reconstruction ranges from 60,000 to 175,000 [80]. A large percent of ACL patients suffer knee osteoarthritis (OA) for a couple of years after receiving the first ACL reconstruction surgeries.

A critical but challenging issue in tendon/ligament tissue engineering is to regenerate a strong bony insertion. The native structure of this tendon/ligament to bone interface consists of collagen fibers, proteoglycan and mineral composition in gradients with many kinds of cells involved [81]. Nowadays, a number of tendon and ligament tissue engineering studies are aiming to rebuild the injured tendon onsite [82 - 85] with the body's own regeneration processes. Among all these studies, BMSCs and ASCs are two kinds of leading stem cells [86]. Tendon stem cells (TSCs), made up of 5% of the all tendon cell populations, are the new candidates for tendon repair due to their special properties in the stem cell criteria: clonogenicity, self-renewal and multi-differentiation potential [87, 88]. Although the TSCs only have been studied for a short period, there have been some evidences of using TSCs as the seeding cells. Most recently, MSCs have also been found in ligament tissues and they may be used as ligament stem cells in ligament repairs [89].

Tendon and Ligament Tissue Engineering Based on Decellularized Matrices

Decellularized tissue matrices have gained significantly attention in regenerative medicine, because the derived scaffolds could maintain the complex structure of the ECM of a specific tissue/organ and preserve most of the original biochemical and biomechanical properties of the tissue.

It has been tested that the tendon matrix is an ideal material for the stem cells for tendon repair as the crucial niche factors which regulate stem cell functions. In an *in vitro* study, the decellularized dog tendons was sliced to 50 µm sections in thickness. BMSCs derived from the same animals were seeded in those single sections and cultured for 2 days. Then 10 cell-seeded sections were bundled together at the ends to form scaffolds, followed by tenogenic differentiation for up to 14 days in centrifuge tubes. The results demonstrated that cells are aligned between the tendon's collagen fibers and the expressed gene MMP13 of tenomodulin (a tenocyte differentiation marker) are up-regulated; but collagen I is down-regulated compared to the BMSCs before seeding [90]. As a further advancement, the rabbit patellar tendons were decellularized to form the engineered tendon matrix (ETM). The ETM powder was acid modified by acetic acid, and has the abilities of film formation and gel formation when it was exposed under UV light and sodium hydroxide solution, respectively. The seeded TSCs expressed more stemness markers (Oct-4, SSEA-1, SSEA-4 and nucleostemin) compared with the cells cultured on plastic surfaces *in vitro*, which means the ETM helps on reserving the undifferentiated state of the stem cells. In addition, the ETM induces TSC proliferation and tenocyte formation. When the TSCs seeded ETM was implanted into the patellar tendons of nude rats for eight weeks, the injured tendons almost restored the organized parallel structure compared with the retained gaps from the stem cells only transplant [91].

Decellularized matrices derived from different tissues have different special organization and alignment thus have their special biological performances on inducing tissue regeneration. The tenogenic effects of decellularized matrices originate from tendon, bone and dermis were compared with nude mouse ectopic tendon formation model and the most suitable one was implanted in a rat Achilles tendon repair model with human TSCs. The 4-week results of mouse ectopic tendon formation demonstrated decellularized tendon matrix was the best candidate for neo-tendon tissue formation: the seeded cells showed spindle-like morphology and aligned parallel to the collagen fibers with the expression of SCX and Eya2 significantly higher than the control groups of decellularized bone and dermis scaffolds. Meanwhile, relatively higher expression of collagen I, III, X, Runx2 and osteocalcin was found in the decellularized bone matrices; strong ALP staining was observed in both decellularized bone and dermis matrices. In other words, the natural collagenous matrix derived from tendon tissue promoted tenogenic-lineage differentiation but inhibited osteogenic behavior of the TSCs, while what derived from bone tissue robustly induced osteogenic-lineage differentiation and what derived from dermis may also help on osteoinduction. Furthermore, when the TSC seeded tendon derived scaffolds were implanted into rat Achilles tendon model with a 6mm gap, the healing achieved larger collagen fiber diameters and stronger stiffness than the cell-free controls [92].

Tendon and Ligament Tissue Engineering Based on Natural Polymers

Silk has been used in clinic for decades with the typical products of sutures, which is also a popular biomaterial in tissue engineering research. The generally used silk is *Bombyx Mori* silk, with silk fibroin (SF) as the main component [93]. Silk or SF solution could form engineered tendon or ligament scaffolds in forms of silk bundles [94, 95], knitted nets [81, 96], and porous sponges [97]. Spider silk is another remarkable natural biomaterial. Spider can make seven different kinds of silk for different use and with different composition. Of all the spider silks, dragline silk of *Nephila clavipes* spider has the strongest tensile strength and super contraction ability. Meanwhile, spider silk shows great tissue compatibility and has been commercialized for five brands of surgical sutures [98]. Nowadays, several strategies to produce recombinant spider silk with bacteria have been developed [99].

With the aim of reconstructing ACL, Shen *et al.* [100] employed knitted silk (*Bombyx mori*) scaffolds and silk-collagen I sponge scaffolds together with primary rabbit MSCs on New Zealand white rabbit models after excising the whole native ACLs. The cells spread out and showed spindle-like morphology on silk-collagen scaffolds with more gene transcription of collagen I, decorin, biglycan and tenascin, compared with cells on silk only scaffolds with tight and stellate morphology. At 18 months post-surgery, the silk scaffold group achieved similar results as silk-collagen group samples showed at 2 months post-surgery (partially covered by regenerated fibrous tissues), but still less collagen were deposited. On the other hand, a typical structure of ligament-bone junction with fibrocartilage and calcified fibrocartilage had formed in the silk-collagen group. The joint surface cartilage of the silk-collagen group was also smoother than the pure silk group.

The same team also explored the same materials on Achilles tendon gap repairing model. The injuries were created as full-thickness, 6 mm long segment on the Achilles tendons of Sprague-Dawley rats. Human ESCs derived MSCs (hESC-MSCs, control group) or the same cells with transferred SCX gene (SCX$^+$hESCs-MSCs, SCX$^+$ group) were seeded in the silk-collagen scaffolds described above. Thereafter, the cell-scaffolds complexes were implanted into the defects. At 2 weeks post-surgery, the SCX$^+$ group had more collagen deposition with more continuous fibers too. In addition, more cells showed spindle-like morphology along the longitudinal axis. Although the SCX$^+$ group achieved significant higher tensile modulus and failure stress in comparison to the controls at 4 weeks and 8 weeks after modeling, the elastic modulus of these constructs are not as high as normal tendons. Considering that the SCX gene also regulates the bone and cartilage formation besides tendon differentiation, the authors examined this with an *in vitro* study and concluded that neo-tendon with SCX$^+$ cells had less calcium nodule formation so that with less osteogenic potential [81].

Alginate is another candidate for tendon regeneration, while dental derived stem cells, including periodontal ligament stem cells (PDLSCs), gingival mesenchymal stem cells (GMSCs) have recently been investigated for tendon reconstruction. To test neo-tendon regeneration feasibility, the PLSCs, GMSCs and BMSCs were encapsulated into RGD coupled alginate hydrogel beads with TGF-β3 ligands, and were subcutaneously transplanted into the nude mice. These results indicated that both PDLSCs and GMSCs advanced over BMSCs, with no osteogenic differentiation was observed and higher levels of tendon markers (SCX, Dcn, Tnmd and Bgy) expressed [101].

Tendon and Ligament Tissue Engineering Based on Synthesized Scaffolds

One reason to use synthesized scaffolds instead of natural polymers is that there is more freedom in designing the structures, biomechanical and biochemical properties for tissue regeneration. The regular methods are modifying the crystallinity, cross-linking, and branching degree of polymers as well as adjusting the molecular weight of precursor or percentage of precursors of copolymers [102].

In treatments of rotator cuff tears, the general solution is to reconnect the tendon to bone using suture. However, bone loss at the repair site and poor scar tissue deposition are the two typical failure responses. To improve tendon-to-bone healing, Lipner *et al.* seeded ASCs with/without BMP-2 in a poly lactic co-glycolic acid (PLGA) scaffold then implanted the scaffold into a rat rotator cuff model. However, the osteogenic factor BMP-2 contained group shows worse results compared to the control group, including delayed healing response, more bone loss, reduced mechanical properties and indicates the BMP-2 is not suitable for the repair [103].

In another study with the aim of Achilles tendon regeneration, the human-induced pluripotent stem cells (hiPS) induced MSCs were seeded onto the chitosan, poly(L-lactic acid) (PLLA), gelatin and poly(-ethylene oxide) (PEO) made scaffold at a mass ratio of 62.1:20.7:10.3:6.9 using stable jet electrospinning (SJES). The *in vitro* results showed that the iPS-MSC preferred to undergo tenogenic differentiation rather than osteogenic differentiation on aligned scaffolds with elevated expression of Col1a1 (>50×), Bgn (>40×), MKX and Fmod (>10×), and some other tendon differentiation markers (SCX, TNC, Tnmd) by more than 2 folds. Meanwhile, the expression of osteogenic marker ALP was suppressed. An animal model was created by removing 6 mm length of Achilles tendon of adult female Sprague-Dawley rats. The cell seeded scaffolds were implanted into the defect area by suturing to the remained tissue with non-resorbable suture soon after defecting. In 2 weeks and 4 weeks post-surgery, more cells spread along the fibers and more collagen was deposited in the aligned scaffolds comparing with those in non-aligned scaffolds. The tenogenic markers were expressed much higher in the aligned scaffolds: Col1a1 (3×), Col5a1(>5×), Bgn(>6×), Fmod (>16.2×) and SCX (8.5×). The diameters of the collagen fibers in aligned scaffold group were found larger than that of non-aligned scaffold group. Additionally, the aligned fibers also achieved higher mechanical properties on stiffness, stress and Young's modulus [84]. Great success of Achilles tendon repair was also achieved on PGA/PLA scaffolds with using ASCs. The scaffold was designed as a cord-net structure with PGA unwoven fibers inside and covered by PGA/PLA fiber net outside, then the ASC-scaffold constructs were implanted into rabbit models with 3 cm defects. At 12 weeks post-surgery, immature tendon-like tissues formed at the defect side. At 45 weeks, the new formed tendon showed shiny-white color and cord-like morphology, with no obvious scaffold residue, and the tensile strength was about 60% of normal tendon tissue [104].

Tendon and Ligament Tissue Engineering Based on Combination of Synthesized and Natural Biomaterials

Silk fibers provide superior mechanical properties and thermal stability but slow degradation time. To reduce their degradation rate to incorporate the regenerated tissue Fang, *et al.* combined PLGA with silk fibers at a mass ratio of 36:64 and weft-knitted into long stripped meshes. Rat tail collagen I was dropped onto the mesh and thereafter, the whole unit was lyophilized and sterilized with ethylene oxide for further use. Rabbit autologous MSCs were cultured on the PLGA/silk-collagen scaffolds for 14 days prior to implanting to the rabbit Achilles tendon defect models (about 2.5 cm in length). The maximum load measured at the rabbit Achilles tendons is almost 2 times higher than the control after 16 weeks. The collagen III mRNA expression was determined after 8 weeks using histology [105]. The previous results could be supported by another *in vitro* study with the same materials, but with knitted *Bombyx mori* silk as the bottom matrix and electrospun PLGA-FGF ultrafine fibers on top. The FGF release could be sustained for over 2 weeks, in a concentration range of 6.5-13.5 pg/ml for the first week. Rabbit BMSCs were seeded on top of the scaffolds. Their results showed that with the chemical stimulation of FGF, the cells were mainly proliferating in the first week, followed by fibroblast differentiation in the second week, and produced more ECM proteins than the polymer only group did [106].

Adjuvant Methods for Tendon and Ligament Regeneration

Mechanical stimulation may accelerate tissue regeneration by activating cells in surrounding host tissues with the processes of surface receptor and ion channel activation and nutrient transport improvement, then finally result in elevated cell proliferation rate and selected differentiation direction. Low intensity stretching increased tenogenic differentiation of TSCs; while high intensity stretching induced the same cells toward adipogenic, osteogenic, and chondrogenic differentiations [107]. When the pre-stretched TSC-seeded poly(L-lactide-c-ε-caprolactone)/collagen (P(LLA-CL)/Col) scaffolds were implanted in rabbit patellar tendon injury models, more mature neo-tendon tissues were observed and the mechanical properties at injured tendons were retained better than in the non-stimulated control group. Faster cell proliferation, more collagen synthesis together with tendon-related ECM protein (Col I, Col III, tenascin C) expression, but down-regulated bone/cartilage formation genes (Col II, Runx2, aggrecan) were observed in the stretching stimulation group as well. This study provided evidence that dynamic mechanical stimulation would be helpful for the maturation of engineered tendons [108].

Platelets are natural reservoirs of growth factors, such as BMPs [109], PDGF, TGF-β1 [110], EGF, hepatocyte growth factors (HGF) [111] and VEGF [112]. Therefore, platelet rich plasma (PRP) has been broadly employed for tissue reconstructions including the treatment of tendon tears in clinic [111]. On rat Achilles tendon injury models, treatments with using collagen sponge and CD44$^+$/CD90$^+$ TSCs with or without PRP have been investigated. The results showed up-regulated Col I/ III, tenascin C and Smad8 expression, which suggested that PRP had the ability of promoting TSCs undergo tenogenic differentiation [109]. However, another study found different platelet concentrations stimulate differently towards tenocytes' behavior: in a proper range of 0.5-1×10^6 plt/μL, the cell proliferation was significantly elevated in 120 h; higher concentrations in 1.0-2.0 ×10^6 plt/μL could also enhance proliferation although to a lower extent; while with platelet concentration higher than this level, the cells tended to form clusters and die. In contrast, MMP production increased with the platelet concentration went up, which meant the excessive high concentration of platelets would lead to decreased tendon mechanical properties [112]. Platelet lysate has similar function on guiding ligament regeneration, which was tested in an *in vitro* study with using 3D-printed polycaprolactone fumarate scaffolds and ASCs. Compared to serum, the platelet lysate induced higher cell proliferation rate and this phenomenon could be further enhanced by FGF-2. In addition, the platelet lysate and FGF-2 combination also increased the expression of tenascin-C and collagen [113].

In summary, advancements in stem cell biology, biomaterials and clinical practice have provided valuable insights in tendon and ligament healing. Matrices composed of natural and/or synthetic biopolymers are critical elements in tissue engineering strategies. Using the tissue engineering matrix as delivery and sustained release vehicles of chemoattractants or therapeutic molecules will enhance cell adhesion, proliferation, migration and differentiation as designed, and ultimately produce a new ECM to facilitate a functional neotissue formation.

SKELETAL MUSCLE TISSUE ENGINEERING

Severe muscle loss and degenerative conditions of the skeletal muscle system still remain as a bottleneck in clinical practice, even though the innate regenerative ability of skeletal muscle is robust because of its tissue-specific stem cell population. Involuntarily, the possibility of musculoskeletal injuries and degenerative conditions keeps increasing when the aging of population continues, which stretches the whole healthcare system physically and financially.

Although tissue grafts have been applied in clinical practice a lot and considered as the gold standard at current stage, their clinical suitability and functionality have been questioned due to slowly spatial remodeling, instability, substandard mechanical performance, *etc.* To address these issues, tissue engineering approaches based on scaffolds, growth factor, and stem cell biology have been employed. Scaffold-based tissue engineering hypothesizes that the exogenous stimuli can induce the self-repairing mechanism of the endogenous cell pools. Briefly, natural or synthetic or hybrid scaffolds are functionalized with biophysical, biochemical, and/or biological signals to achieve spatial cues, stem cell proliferation, migration, and differentiation, controlled delivery and release of therapeutic molecules. Promising results have been reported in both preclinical and clinical settings. Meanwhile, scaffold-free therapies have also attracted lots of research attention and been reviewed in several great publications.

Satellite Cell and Its Niche

As mentioned above, the extraordinary self-healing ability of skeletal muscle system comes from its own tissue-special stem cells: satellite cells. The satellite cells (SCs) locate beneath the basal membrane and next to the sarcolemma, in some vascular rich regions where they could interact with other muscle resident stem/progenitor cells *via* paracrine [114]. The primary surface and nuclear biomarkers of the SCs are paired box transcription factors Pax7 and Pax3, however, these two factors may not be positive simultaneously [115]. Pax7 is expressed by the canonical SCs in both G0 phase (quiescent stage) and proliferating stage. Pax 3 only associates with muscles like diaphragm and trunk muscles. Satellite cells can be sorted into a heterogeneous population based on biomarkers: SM/C-2.6, Integrin, CXCR4, CD34 and non-satellite cell markers: CD31, CCD45, Sca-1, Mac-1. Among all the satellite cells, the Pax7[+] population have showed outstanding regenerative potential, that only 900 cells generated new muscles with measurable contractions [116]. Many other stem cells also enable skeletal muscle regeneration, such as MSCs [117], muscle derived stem cells [118], muscle-derived CD133[+] stem cells [119], mesoangioblasts [120], ESCs and iPSCs [121 - 123]. However, the SCs are still the most promising one at present.

Similar to other stem cells, SCs also reside in their unique niche microenvironment, and start losing their regenerative potential after being isolated. The ECM composition, topography, stiffness and porosity are the main characters of the SC nice, which affect SC activation, adhesion, proliferation, migration, differentiation and communication with other cells [114]. Among all the ECM molecules, laminin plays a more critical role regarding stem cell proliferation and myogenic differentiation compared to collagen I, gelatin and fibronectin [124]. Physical exercise and mechanical loading remodels the skeletal muscle ECM, while age related ECM degeneration could not be turned over [114]. Therefore, skeletal muscle tissue engineering is looking forward to building specific synthetic niches to support muscle therapeutic stem cells and retain their innate capabilities with 3D biomaterial scaffolds and growth factors.

Skeletal Muscle Tissue Engineering Based on Decellularized Muscle Derived Extracellular Matrix

Because the skeletal muscle ECM decides stem cell fate to a large extent, decellularized muscle derived ECM (mdECM) is considered as the best option to reconstruct skeletal muscle due to its maintained nature structure and molecules. Porzionato compared mdECM collected from human amputed limbs and cadavers, with two different cell removing protocols. Both muscle resources had rich collagen, elastic fibers, GAG and proteoglycan, showed no significant difference on ultra-structures as well. However, trypsin-EDTA/Triton X-NH$_4$OH combination worked more effectively on decellularization than sodium deoxycholate and DNase I. When the cell free mdECM grafts were implanted in rabbit model, fibroblast invasion and fibrous connective tissue formation was observed around the interface of graft-host tissue, but not in the center of the matrix. The mdECM showed acceptable results with respect to mechanical properties and biointegration [125]. On the 3-D bioprinting platform, Choi *et al.* prepared a kind of temperature sensitive bioink by dissolving mdECM in acidic solution with pepsin. This bioink showed great printability for shaping volumetric variable architectures under control. Encapsulated C2c6 myoblasts displayed high viability, proliferation and myotube formation abilities and successfully underwent myogenic differentiation [126].

Skeletal Muscle Tissue Engineering Based on Natural Materials

The family of collagen exist abundantly within the ECM in connective tissues, especially the type I collagen [127]. The regenerative applications of collagen scaffolds have been investigated in a wide variety of tissues, such like skin, cartilage, bone, tendon, skeletal muscle, and nerve [127, 128]. In skeletal muscle regeneration, collagen scaffolds have been utilized in mainly two ways: one is as cell delivery vehicles and the other is as the sustained release buffer of growth factors.

As cell delivery vehicles, the collagen scaffolds can temporarily fill the vacancy due to the ECM loss and provide supports for regenerative processes, especially the integration to the host myofibers [129 - 131]. Because of high aligned arrangement of natural myofibers, collagen scaffolds with regulated and aligned pore structure are preferred. Kroehne *et al.* successfully prepared a collagen I sponge with parallel pore structure between 20 - 50 μm. These artificial scaffolds were infiltrated with C2c6 cells (a permanent myogenic cell line), incubated firstly in expansion medium for myoblasts, then in fusion medium to induce differentiation and fusion of myotubes, which resulted in a parallel arrangement of myotubes along the pore structure. Afterwards, they transplanted these collagen sponges with either proliferated cell population or with fused myotubes into the beds of excised anterior tibial muscles of immunodeficient host mice, which were genetic modified for enhanced green fluorescent protein (eGFP) in order to separate host contributed muscle tissue from grafted ones. Histological analysis showed that the donor muscle fibers integrated into the host tissues in the outer area of grafts. Mechanical forces were also confirmed from regenerated muscle tissues by directly electrical stimulations [132].

The other way of utilizing collagen scaffolds is the chemically modification to achieve controlled release of growth factors into the injury site [133, 134], which can significantly benefit the muscle regeneration process by biochemical stimulation. For example, VEGF-loaded collagen scaffold can benefit the muscle functional recovery after acute ischemia from 53% to 75% of the force production from natural muscles [135]. Meanwhile, HGF, IGF-1, and FGF-2 have also been employed to attract stem cells into the injury sites for higher regenerative potential [136, 137].

Besides collagen scaffolds, alginate derived scaffolds are becoming more and more popular in the area of skeletal muscle regeneration. Although extracted from seaweed, alginate gains its own position in skeletal muscle regeneration because of its capability of chemical modification, tunable mechanical properties, and well-controlled structures [138 - 141]. Mooney's group has reported an optimized stiffness of alginate hydrogels for myoblast proliferation and differentiation, which was 13 to 45 kPa [142]. Meanwhile, cell adhesion onto alginate scaffolds modified with RGD binding domain also regulates myoblasts' proliferation and differentiation [143, 144]. In addition, alginate scaffolds could gain controlled release ability of various growth factors through chemical modification and crosslinking, which are important to stimulate stem cells to enter regenerative stages. *In vitro* studies showed that HGF and FGF2 significantly boosted the viability of myoblast within alginate scaffolds over a 5-day course and stimulated outward migration from scaffolds into host tissues to 110% of the original cell number [145]. *In vivo* studies on a mouse tibialis anterior laceration model showed that alginate scaffolds with HGF, FGF-2, and myoblasts significantly reduced the wound size at 30 days post-surgery, and regenerated new myofibers. However, without combination, each growth factor could only deliver a modest regeneration [146]. In ischemic skeletal muscle injuries, IGF-1 and VEGF are another pair of growth factors that have been widely investigated [149 - 153]. It was believed that in combination of alginate, IGF-1 and VEGF could induce early vascularization and myoblast survival within the injury sites [149, 151, 153] Therefore, co-delivery of IGF-1 and VEGF within alginate scaffolds were able to increase the force production 4 to 5-fold of the black alginate gels at 7 weeks post-surgery, which got closing to the performance of the normal tissues [147].

Other natural biopolymers are also investigated for their potentials in skeletal muscle regeneration. Zhang *et al.* developed a chitosan scaffold with proper stiffness and uniaxial pore structure from temperature gradient, whose pore diameters controllably decreased with increased temperature gradient [148]. Myogenic cells (C2c6) were seeded into preformed chitosan scaffolds and cultured for two weeks in standard culture media without any growth factors. According to the authors, myotubes matured in this chitosan scaffolds had mean myotube diameters 3–5 times greater than various studied uniaxial materials for skeletal muscle regeneration, and the lengths of the myotubes were mainly determined by culture time if through-channels inside the scaffolds were not interrupted. Coppi *etf al.* developed an *in situ* photo-cross-linkable HA–PI scaffold by conjugating hyaluronic acid (HA) and 1-[4-(2-hydroxyethoxy)-phenyl]-2-hydroxy-2-methyl-1-propane-1-one (PI) as photoinitiator [149]. Extracted satellite cells were encapsulated into hydrogel and implanted into partially ablated tibialis anterior of C57BL/6J mice. The results showed a promising promotion in muscle structure, number of regenerated myofibers, and neural and vascular networks, as well as functional recovery with assessed contraction force. Wang *et al.* developed a gelatin-hydroxyphenylpropionic acid (Gtn-HPA) hydrogel [150]. They reported that on a stiff hydrogel (G' > 8 kPa), hMSCs showed higher upregulated myogenic markers, which proved another route to achieve myogenesis through hMSCs and biophysical properties of the scaffolds. While novel natural biopolymer scaffolds have been continuously developed for skeletal muscle generation, macroscale alignment of pore structure, degradation rate, immunoresponse, optimized combination of biophysical parameters and biochemical stimulations, and clinical trials remain as major challenges for all natural biomaterials.

Skeletal Muscle Tissue Engineering Based on Synthetic Materials

Comparing with natural materials, synthetic biopolymers provide precisely controlled physical and chemical properties, as well as easily grafted cell attachment moieties like RGD and controlled release vehicles for growth factors [151]. A broad range of synthetic biopolymers have been investigated for myogenesis, including poly(ε-caprolactone) (PCL) [152], polyurethanes (PU) [153, 154], poly(lactic acid) (PLA) [155], poly(glycolic acid) (PGA) [156, 157], and their copolymers [158]. Ready in fabrication is a great advantage for synthetic polymers that alignment, topography, and functionalization can be incorporated easily. Porous PLGA scaffolds have been prepared and proved to promote neovascularization and cell migration *in vivo* [159 - 161]. Several porous polyurethane scaffolds were prepared and implanted in the rat abdomen and composition of PU scaffolds had significantly altered stiffness change, degradation kinetics, and remodeling process [162]. Electrospun PCL meshes promoted the alignment of seeded myoblasts along the fibers [163], as well as with a grooved topography that also increased the length of myotubes [164]. Conductive polyaniline (PANI) was intentionally blended in PCL and aligned nanofiber scaffolds were prepared by electrospinning. As-prepared conductive scaffolds provided a more efficient electrical stimulation during myoblast culturing, which was found to strongly promote myoblast differentiation comparing with aligned PCL fiber scaffolds alone [165, 166].

Although these synthetic scaffolds have succeeded in their tailorable mechanical property, ease in fabrication, and incorporated with growth factors to solve cell attachment issues, foreign body reaction (FBR) including inflammation, fibrous capsule formation and protein adsorption still need to be taken care of, some synthetic materials are even considered as non-degradable [167 - 170]. Based on the limited knowledge of host-material interaction, it is difficult to predict the extent of FBR to the surface morphology, chemical and physical properties of a new biomaterial [171].

Skeletal Muscle Tissue Engineering Based on Hybrid Materials

Both types of scaffolds exhibit their own advantages in skeletal muscle regeneration, as well as inherent limitations. Natural materials are usually not as good as synthetic materials in mechanical stability, nor readily in fabrication for aligned microstructure or topography. While synthetic materials are easily failed in issues like lack of cell adhesion molecules and biocompatibility, immune response irritation, or transplant rejection. Straightforwardly, hybrid materials possess both natural and synthetic materials have been developed. The synthetic component provides elasticity, mechanical support, and sometimes alignment to guide myoblasts, while natural materials are responsible for cell adhesion, proliferation, migration, differentiation, neovascularization, and fusion with host tissues. In general, several configurations of hybrid materials are specifically interested in skeletal muscle regeneration, such like encapsulating fibers/meshes in a hydrogel matrix, natural materials coating on synthetic meshes, and electrospinning of hybrid composites [172].

For example, polypropylene (PP) fibers reinforced acellular dermal ECM could increase mechanical strength of the ECM grafts, prevent premature failure, and contribute to the durability of reinforced ECM towards enzymatic degradation *in vitro* [173]. PANI and poly(3,4-ethylenedioxythiophene) (PEDOT) conductive polymer nanofibers were embedded in Type I collagen hydrogel to fabricate very special conductive scaffolds without compromising cell viability, proliferation or differentiation potentials [174]. As-prepared scaffolds were also injectable for minimal invasion procedure and increased electrical conductivity of the scaffolds by more than 400% comparing with collagen controls, which supported its great potential in nerve and skeletal muscle regeneration.

PCL/chitosan electronspun scaffolds possessed aligned nanofibers and provided better myoblast alignment than unaligned PCL/chitosan films. By adding chitosan microfiber bands along the same alignment of electronspun scaffolds, myoblast differentiation was further boosted and skeletal muscle myosin heavy chains (MHC) were expressed [175, 176]. In another study, polyurethane elastomer and

ECM hydrogel were concurrent electrospun to form a sandwich structure because component ratio could be varied continuously during the fabrication. The top and base layers were designed to provide mechanical support and flexibility thus were rich in PU component, while the mid layer was proposed for cell immigration thus was rich in ECM component [177, 178]. Type I collagen was also employed in electrospinning with PCL to build scaffolds and the *in vitro* results suggested enhanced cell viability, cell adherence, myoblasts alignment, and myofiber fusion in comparison with PCL scaffolds alone [179, 180]. Similarly, gelatin was blended into PCL and eletrospun into scaffolds, then genipin was employed to cross-link as-prepared nanofibers in order to enhance its mechanical performance [181]. As a result, myotube formation could be found only on PCL-gelatin nanofibers, and the MHC expression was promoted when the ratio of gelatin content increased.

In the surface coating configuration, natural materials are not required to have its own microstructure or form hydrogels, but should adhere onto the surface of the synthetic materials by physical interaction or chemical conjugation. Polyurethane microchannel scaffolds coated by chemically cross-linked gelatin or silk fibroin were reported to enhance myotube formation [182]. Polypropylene meshes, which were coated with digested ECM solution by simply soaking and air drying, could reduce the foreign body response, and the coating was robust enough to withstand numerous washes with saline [183]. Lesman *et al.* prepared a PLLA/PLGA sponge soaking coated by fibrin hydrogel and tri-cultured them with endothelial cells, fibroblasts and tissue specific skeletal myoblast cells. The complex scaffolds showed significant enhancement in vascularization inside the network, in which fast remodeled fibrin construct provided bioactivity while the PLLA/PLGA sponge provided mechanical stability [182]. Additionally, cellulose [184, 185], omega-3 fatty acids [186, 187], and hyaluronic acid [100, 179] have also been investigated for surface coating applications onto synthetic meshes and boost their biological performances.

Vascularized Skeletal Muscle Construction

Angiogenesis and vascularization is a principle obstacle for skeletal muscle reconstruction, especially for large volumetric defects, because sufficient oxygen and nutrition supply is necessary for long-term survival of the engineered tissue. VEGF is a signal protein trigger angiogenesis and vessel formation during embryonic development, injured tissue repair, and exercise induced skeletal muscle vascularization. With SDF-1α, VEGF showed synergistic effects on skeletal muscle regeneration *in vivo* [188]. Most recently, epithelial-mesenchymal transition was also testified as mainly dependent on VEGF *via* PI3K/Akt/mTOR pathway [189]. Hypoxia treatment and low dose laser exposure are also considered as useful strategies in vascularized skeletal muscle engineering [190]. Macrophages (MPs) infiltration benefits vascular remodeling in the injured muscle by secreting pro-angiogenic growth factors which promotes endothelial-derived progenitors undergo neo-capillary formation [191].

Several approaches of engineering vascularized skeletal muscle have been established with both natural polymers and hybrid materials. With combination of human VEGF-165 and SDF-1 gene transferred skeletal myoblasts derived from newborn rats and bovine collagen sponge scaffolds, Zhou *et al.* constructed a muscle repair patch and sutured it in the dorsal muscle defect area in a rat wound healing model. Eight weeks postsurgery, the cell laden scaffold was partly absorbed with non-clear boundaries between the host tissue, and was covered by the fascia. The artificial tissue seemed well-vascularized, similar to the native tissue with no observable scar formation [188]. Within similar animal model, Juhas *et al.* implanted $Pax7^+/MyoD^+$ SCs laden Matrigel-based muscle bundles in the dorsal muscle of rats. During 2-week healing period, vascular formation and blood perfusion was gradually observed. The engineered muscle grafts exhibited robust myogenesis and self-regeneration, with intracellular calcium handling and contractile function. This research successfully built an *in vitro* platform for drug screening of skeletal muscle diseases [192, 193].

In summary, skeletal muscle regeneration has gained significant progresses because of the development of tissue regeneration biomaterials, especially the hybrid scaffold materials. The development of scaffolds should aim on microstructure that can guide myotube formation, as well as artificial stem cell niches to proliferate and differentiate therapeutic stem cells, either endogeneous or exogeneous. Furthermore, it is necessary to achieve clinically relevant sizes and shapes as well as increase the clinical feasibility when applying stem cells/scaffolds in future research.

CONCLUSION AND FUTURE PERSPECTIVES

Musculoskeletal system tissue engineering is one of the classic research topics. The regeneration strategies based on scaffolds and stem cells have experienced significant development in the past two decades, with many technologies ready to be translated to clinical applications. Recent work with highlights of adopting new technologies, such as electrospinning, 3-D bioprinting, and micro/nano drug delivery system, will bring the work with traditional biomaterials to a higher level. Meanwhile, the development of new biomaterials, or new modification methods of well-studied materials, opens new research fields for tissue engineering studies. As a series of biological or biomedical engineering research achievement has been applied in orthopedic clinics, such as extracorporeal shock wave therapy and electric stimulation therapy, the regenerative treatment outcomes will be enhanced in the future when we involve these new skills.

ABBREVIATIONS

ACL	anterior cruciate ligament
ASCs	adipose derived stem cells
ECM	extracellular matrix
EGF	epidermal growth factor
ESCs	embryonic stem cells
ETM	engineered tendon matrix
iPSCs	induced pluripotent stem cells
IGF	insulin growth factor
MHC	myosin heavy chains
MSC	mesenchymal Stem cells
PANI	polyaniline
PCL	polycaprolactone
PET	polyethylene terephthalate
PVA	polyvinyl alcohol
PGA	polyglycolic acid
PLGA	polylactic-*co*-glycolic acid
PRP	platelet rich plasma
TMSCs	turbinate mesenchymal stem cells
TSCs	tendon stem cells
VEGF	vascular endothelial growth factor

CONFLICT OF INTEREST

The authors declare no conflict of interest, financial or otherwise.

ACKNOWLEDGEMENTS

Declared none.

REFERENCES

[1] Kimelman N, Pelled G, Helm GA, Huard J, Schwarz EM, Gazit D. Review: gene- and stem cell-based therapeutics for bone regeneration and repair. Tissue Eng 2007; 13(6): 1135-50.
[http://dx.doi.org/10.1089/ten.2007.0096] [PMID: 17516852]

[2] Mauney JR, Volloch V, Kaplan DL. Role of adult mesenchymal stem cells in bone tissue engineering applications: current status and future prospects. Tissue Eng 2005; 11(5-6): 787-802.
[http://dx.doi.org/10.1089/ten.2005.11.787] [PMID: 15998219]

[3]   Meyer U, Wiesmann HP, Berr K, Kübler NR, Handschel J. Cell-based bone reconstruction therapies-principles of clinical approaches. Int J Oral Maxillofac Implants 2006; 21(6): 899-906.
        [PMID: 17190299]

[4]   Pioletti DP, Montjovent MO, Zambelli PY, Applegate L. Bone tissue engineering using foetal cell therapy. Swiss Med Wkly 2006; 136(35-36): 557-60.
        [PMID: 17043947]

[5]   Bianco P, Robey PG. Stem cells in tissue engineering. Nature 2001; 414(6859): 118-21.
        [http://dx.doi.org/10.1038/35102181] [PMID: 11689957]

[6]   Bersenev Alexey. Trends in cell therapy clinical trials 2011–2014. Cell Trials blog 2015 February 14; Available: http://celltrials.info/2015/02/14/trends-2014/

[7]   Murphy WL, McDevitt TC, Engler AJ. Materials as stem cell regulators. Nat Mater 2014; 13(6): 547-57.
        [http://dx.doi.org/10.1038/nmat3937] [PMID: 24845994]

[8]   Tollemar V, Collier ZJ, Mohammed MK, Lee MJ, Ameer GA, Reid RR. Stem cells, growth factors and scaffolds in craniofacial regenerative medicine. Genes Dis 2016; 3(1): 56-71.
        [http://dx.doi.org/10.1016/j.gendis.2015.09.004] [PMID: 27239485]

[9]   Luu HH, Song W-X, Luo X, *et al.* Distinct roles of bone morphogenetic proteins in osteogenic differentiation of mesenchymal stem cells. J Orthop Res 2007; 25(5): 665-77.
        [http://dx.doi.org/10.1002/jor.20359] [PMID: 17290432]

[10]   Ishikawa H, Kitoh H, Sugiura F, Ishiguro N. The effect of recombinant human bone morphogenetic protein-2 on the osteogenic potential of rat mesenchymal stem cells after several passages. Acta Orthop 2007; 78(2): 285-92.
        [http://dx.doi.org/10.1080/17453670710013816] [PMID: 17464620]

[11]   Baugé C, Boumédiene K. Use of adult stem cells for cartilage tissue engineering: current status and future developments. Stem Cells Int 2015; 2015: 438026.
        [http://dx.doi.org/10.1155/2015/438026]

[12]   Friedenstein AJ, Piatetzky-Shapiro II, Petrakova KV. Osteogenesis in transplants of bone marrow cells. J Embryol Exp Morphol 1966; 16(3): 381-90.
        [PMID: 5336210]

[13]   Owen M. Marrow stromal stem cells. J Cell Sci Suppl 1988; 10: 63-76.
        [http://dx.doi.org/10.1242/jcs.1988.Supplement_10.5] [PMID: 3077943]

[14]   Kadiyala S, Jaiswal N, Bruder SP. Culture-expanded, bone marrow-derived mesenchymal stem cells can regenerate a critical-sized segmental bone defect. Tissue Eng 1997; 3(2): 173-85.
        [http://dx.doi.org/10.1089/ten.1997.3.173]

[15]   Meinel L, Karageorgiou V, Fajardo R, *et al.* Bone tissue engineering using human mesenchymal stem cells: effects of scaffold material and medium flow. Ann Biomed Eng 2004; 32(1): 112-22.
        [http://dx.doi.org/10.1023/B:ABME.0000007796.48329.b4] [PMID: 14964727]

[16]   Kon E, Muraglia A, Corsi A, *et al.* Autologous bone marrow stromal cells loaded onto porous hydroxyapatite ceramic accelerate bone repair in critical-size defects of sheep long bones. J Biomed Mater Res 2000; 49(3): 328-37.
        [http://dx.doi.org/10.1002/(SICI)1097-4636(20000305)49:3<328::AID-JBM5>3.0.CO;2-Q] [PMID: 10602065]

[17]   McIntosh K, Zvonic S, Garrett S, *et al.* The immunogenicity of human adipose-derived cells: temporal changes *in vitro.* Stem Cells 2006; 24(5): 1246-53.
        [http://dx.doi.org/10.1634/stemcells.2005-0235] [PMID: 16410391]

[18]   Zuk PA, Zhu M, Mizuno H, *et al.* Multilineage cells from human adipose tissue: implications for cell-based therapies. Tissue Eng 2001; 7(2): 211-28.

[http://dx.doi.org/10.1089/107632701300062859] [PMID: 11304456]

[19]   Gronthos S, Mankani M, Brahim J, Robey PG, Shi S. Postnatal human dental pulp stem cells (DPSCs) *in vitro* and *in vivo*. Proc Natl Acad Sci USA 2000; 97(25): 13625-30.
[http://dx.doi.org/10.1073/pnas.240309797] [PMID: 11087820]

[20]   Gronthos S, Brahim J, Li W, *et al.* Stem cell properties of human dental pulp stem cells. J Dent Res 2002; 81(8): 531-5.
[http://dx.doi.org/10.1177/154405910208100806] [PMID: 12147742]

[21]   Lee OK, Kuo TK, Chen WM, Lee KD, Hsieh SL, Chen TH. Isolation of multipotent mesenchymal stem cells from umbilical cord blood. Blood 2004; 103(5): 1669-75.
[http://dx.doi.org/10.1182/blood-2003-05-1670] [PMID: 14576065]

[22]   Bieback K, Kern S, Klüter H, Eichler H. Critical parameters for the isolation of mesenchymal stem cells from umbilical cord blood. Stem Cells 2004; 22(4): 625-34.
[http://dx.doi.org/10.1634/stemcells.22-4-625] [PMID: 15277708]

[23]   Sarugaser R, Lickorish D, Baksh D, Hosseini MM, Davies JE. Human umbilical cord perivascular (HUCPV) cells: a source of mesenchymal progenitors. Stem Cells 2005; 23(2): 220-9.
[http://dx.doi.org/10.1634/stemcells.2004-0166] [PMID: 15671145]

[24]   Quarto R, Mastrogiacomo M, Cancedda R, *et al.* Repair of large bone defects with the use of autologous bone marrow stromal cells. N Engl J Med 2001; 344(5): 385-6.
[http://dx.doi.org/10.1056/NEJM200102013440516] [PMID: 11195802]

[25]   Tseng PC, Young TH, Wang TM, Peng HW, Hou SM, Yen ML. Spontaneous osteogenesis of MSCs cultured on 3D microcarriers through alteration of cytoskeletal tension. Biomaterials 2012; 33(2): 556-64.
[http://dx.doi.org/10.1016/j.biomaterials.2011.09.090] [PMID: 22024363]

[26]   Gandavarapu NR, Alge DL, Anseth KS. Osteogenic differentiation of human mesenchymal stem cells on α5 integrin binding peptide hydrogels is dependent on substrate elasticity. Biomater Sci 2014; 2(3): 352-61.
[http://dx.doi.org/10.1039/C3BM60149H] [PMID: 24660057]

[27]   Li X, Huang Y, Zheng L, *et al.* Effect of substrate stiffness on the functions of rat bone marrow and adipose tissue derived mesenchymal stem cells *in vitro*. J Biomed Mater Res A 2014; 102(4): 1092-101.
[http://dx.doi.org/10.1002/jbm.a.34774] [PMID: 23630099]

[28]   Kwon JS, Kim SW, Kwon DY, *et al. In vivo* osteogenic differentiation of human turbinate mesenchymal stem cells in an injectable *in situ*-forming hydrogel. Biomaterials 2014; 35(20): 5337-46.
[http://dx.doi.org/10.1016/j.biomaterials.2014.03.045] [PMID: 24720878]

[29]   Das S, Pati F, Choi YJ, *et al.* Bioprintable, cell-laden silk fibroin-gelatin hydrogel supporting multilineage differentiation of stem cells for fabrication of three-dimensional tissue constructs. Acta Biomater 2015; 11: 233-46.
[http://dx.doi.org/10.1016/j.actbio.2014.09.023] [PMID: 25242654]

[30]   Park JY, Shim JH, Choi SA. 3D printing technology to control BMP-2 and VEGF delivery spatially and temporally to promote large-volume bone regeneration. J Mater Chem B Mater Biol Med 2015; 3: 5415-25.
[http://dx.doi.org/10.1039/C5TB00637F]

[31]   Cui H, Zhu W, Nowicki M, Zhou X, Khademhosseini A, Zhang LG. Hierarchical Fabrication of Engineered Vascularized Bone Biphasic Constructs *via* Dual 3D Bioprinting: Integrating Regional Bioactive Factors into Architectural Design. Adv Healthc Mater 2016; 5(17): 2174-81.
[http://dx.doi.org/10.1002/adhm.201600505] [PMID: 27383032]

[32]   Riew KD, Wright NM, Cheng S, Avioli LV, Lou J. Induction of bone formation using a recombinant

adenoviral vector carrying the human BMP-2 gene in a rabbit spinal fusion model. Calcif Tissue Int 1998; 63(4): 357-60.
[http://dx.doi.org/10.1007/s002239900540] [PMID: 9744997]

[33] Kang Q, Song W-X, Luo Q, *et al.* A comprehensive analysis of the dual roles of BMPs in regulating adipogenic and osteogenic differentiation of mesenchymal progenitor cells. Stem Cells Dev 2009; 18(4): 545-59.
[http://dx.doi.org/10.1089/scd.2008.0130] [PMID: 18616389]

[34] Cipriano AF, Sallee A, Guan RG, *et al.* Investigation of magnesium-zinc-calcium alloys and bone marrow derived mesenchymal stem cell response in direct culture. Acta Biomater 2015; 12: 298-321.
[http://dx.doi.org/10.1016/j.actbio.2014.10.018] [PMID: 25449917]

[35] Antunes JC, Tsaryk R, Gonçalves RM, *et al.* Poly(γ-Glutamic Acid) as an Exogenous Promoter of Chondrogenic Differentiation of Human Mesenchymal Stem/Stromal Cells. Tissue Eng Part A 2015; 21(11-12): 1869-85.
[http://dx.doi.org/10.1089/ten.tea.2014.0386] [PMID: 25760236]

[36] Shirasawa S, Sekiya I, Sakaguchi Y, Yagishita K, Ichinose S, Muneta T. *In vitro* chondrogenesis of human synovium-derived mesenchymal stem cells: optimal condition and comparison with bone marrow-derived cells. J Cell Biochem 2006; 97(1): 84-97.
[http://dx.doi.org/10.1002/jcb.20546] [PMID: 16088956]

[37] Bernhard JC, Vunjak-Novakovic G. Should we use cells, biomaterials, or tissue engineering for cartilage regeneration? Stem Cell Res Ther 2016; 7(1): 56.
[http://dx.doi.org/10.1186/s13287-016-0314-3] [PMID: 27089917]

[38] Madry H, Grün UW, Knutsen G. Cartilage repair and joint preservation: medical and surgical treatment options. Dtsch Arztebl Int 2011; 108(40): 669-77.
[PMID: 22114626]

[39] Brittberg M, Lindahl A, Nilsson A, Ohlsson C, Isaksson O, Peterson L. Treatment of deep cartilage defects in the knee with autologous chondrocyte transplantation. N Engl J Med 1994; 331(14): 889-95.
[http://dx.doi.org/10.1056/NEJM199410063311401] [PMID: 8078550]

[40] Hwang Y-S, Cho J, Tay F, *et al.* The use of murine embryonic stem cells, alginate encapsulation, and rotary microgravity bioreactor in bone tissue engineering. Biomaterials 2009; 30(4): 499-507.
[http://dx.doi.org/10.1016/j.biomaterials.2008.07.028] [PMID: 18977027]

[41] Marolt D, Campos IM, Bhumiratana S, *et al.* Engineering bone tissue from human embryonic stem cells. Proc Natl Acad Sci USA 2012; 109(22): 8705-9.
[http://dx.doi.org/10.1073/pnas.1201830109] [PMID: 22586099]

[42] Wang Y, Yu X, Baker C, Murphy WL, McDevitt TC. Mineral particles modulate osteo-chondrogenic differentiation of embryonic stem cell aggregates. Acta Biomater 2016; 29: 42-51.
[http://dx.doi.org/10.1016/j.actbio.2015.10.039] [PMID: 26597546]

[43] Vonk LA, de Windt TS, Slaper-Cortenbach IC, Saris DB. Autologous, allogeneic, induced pluripotent stem cell or a combination stem cell therapy? Where are we headed in cartilage repair and why: a concise review. Stem Cell Res Ther 2015; 6: 94.
[http://dx.doi.org/10.1186/s13287-015-0086-1] [PMID: 25976213]

[44] Mardones R, Jofré CM, Minguell JJ. Cell therapy and tissue engineering approaches for cartilage repair and/or regeneration. Int J Stem Cells 2015; 8(1): 48-53.
[http://dx.doi.org/10.15283/ijsc.2015.8.1.48] [PMID: 26019754]

[45] Takahashi K, Tanabe K, Ohnuki M, *et al.* Induction of pluripotent stem cells from adult human fibroblasts by defined factors. Cell 2007; 131(5): 861-72.
[http://dx.doi.org/10.1016/j.cell.2007.11.019] [PMID: 18035408]

[46] Wakitani S, Nawata M, Tensho K, Okabe T, Machida H, Ohgushi H. Repair of articular cartilage defects in the patello-femoral joint with autologous bone marrow mesenchymal cell transplantation:

three case reports involving nine defects in five knees. J Tissue Eng Regen Med 2007; 1(1): 74-9.
[http://dx.doi.org/10.1002/term.8] [PMID: 18038395]

[47] Huey DJ, Hu JC, Athanasiou KA. Unlike bone, cartilage regeneration remains elusive. Science 2012; 338(6109): 917-21.
[http://dx.doi.org/10.1126/science.1222454] [PMID: 23161992]

[48] Wang L, Zhang C, Li C, *et al.* Injectable calcium phosphate with hydrogel fibers encapsulating induced pluripotent, dental pulp and bone marrow stem cells for bone repair. Mater Sci Eng C 2016; 69: 1125-36.
[http://dx.doi.org/10.1016/j.msec.2016.08.019] [PMID: 27612810]

[49] de Peppo GM, Marcos-Campos I, Kahler DJ, *et al.* Engineering bone tissue substitutes from human induced pluripotent stem cells. Proc Natl Acad Sci USA 2013; 110(21): 8680-5.
[http://dx.doi.org/10.1073/pnas.1301190110] [PMID: 23653480]

[50] de Peppo GM, Vunjak-Novakovic G, Marolt D. Cultivation of human bone-like tissue from pluripotent stem cell-derived osteogenic progenitors in perfusion bioreactors. Methods Mol Biol 2014; 1202: 173-84.
[http://dx.doi.org/10.1007/7651_2013_52] [PMID: 24281874]

[51] Jin GZ, Kim TH, Kim JH, *et al.* Bone tissue engineering of induced pluripotent stem cells cultured with macrochanneled polymer scaffold. J Biomed Mater Res A 2013; 101(5): 1283-91.
[http://dx.doi.org/10.1002/jbm.a.34425] [PMID: 23065721]

[52] Ratajczak MZ, Bujko K, Wojakowski W. Stem cells and clinical practice: new advances and challenges at the time of emerging problems with induced pluripotent stem cell therapies. Pol Arch Med Wewn 2016; 126(11): 879-90.
[http://dx.doi.org/10.20452/pamw.3644] [PMID: 27906881]

[53] Green WT Jr. Articular cartilage repair. Behavior of rabbit chondrocytes during tissue culture and subsequent allografting. Clin Orthop Relat Res 1977; (124): 237-50.
[PMID: 598084]

[54] LC R. Repair of partial-thickness defects in articular cartilage: cell recruitment from the synovial membrane. J Bone Jt Surg 1996; 78(5): 721-33.
[http://dx.doi.org/10.2106/00004623-199605000-00012]

[55] Kuo CK, Li WJ, Mauck RL, Tuan RS. Cartilage tissue engineering: its potential and uses. Curr Opin Rheumatol 2006; 18(1): 64-73.
[http://dx.doi.org/10.1097/01.bor.0000198005.88568.df] [PMID: 16344621]

[56] Martin I, Baldomero H, Bocelli-Tyndall C, *et al.* The survey on cellular and engineered tissue therapies in Europe in 2011. Tissue Eng Part A 2014; 20(3-4): 842-53.
[PMID: 24090467]

[57] Ferguson CM, Schwarz EM, Reynolds PR, Puzas JE, Rosier RN, O'Keefe RJ. Smad2 and 3 mediate transforming growth factor-beta1-induced inhibition of chondrocyte maturation. Endocrinology 2000; 141(12): 4728-35.
[http://dx.doi.org/10.1210/endo.141.12.7848] [PMID: 11108288]

[58] Lieb E, Vogel T, Milz S, Dauner M, Schulz MB. Effects of transforming growth factor beta1 on bonelike tissue formation in three-dimensional cell culture. II: Osteoblastic differentiation. Tissue Eng 2004; 10(9-10): 1414-25.
[http://dx.doi.org/10.1089/ten.2004.10.1414] [PMID: 15588401]

[59] Mehlhorn AT, Schmal H, Kaiser S, *et al.* Mesenchymal stem cells maintain TGF-beta-mediated chondrogenic phenotype in alginate bead culture. Tissue Eng 2006; 12(6): 1393-403.
[http://dx.doi.org/10.1089/ten.2006.12.1393] [PMID: 16846338]

[60] Panadero JA, Lanceros-Mendez S, Ribelles JL. Differentiation of mesenchymal stem cells for cartilage tissue engineering: Individual and synergetic effects of three-dimensional environment and mechanical

loading. Acta Biomater 2016; 33: 1-12.
[http://dx.doi.org/10.1016/j.actbio.2016.01.037] [PMID: 26826532]

[61]   Silva R, Fabry B, Boccaccini AR. Fibrous protein-based hydrogels for cell encapsulation. Biomaterials
       2014; 35(25): 6727-38.
       [http://dx.doi.org/10.1016/j.biomaterials.2014.04.078] [PMID: 24836951]

[62]   Munarin F, Petrini P, Bozzini S, Tanzi MC. New perspectives in cell delivery systems for tissue
       regeneration: natural-derived injectable hydrogels. J Appl Biomater Funct Mater 2012; 10(2): 67-81.
       [http://dx.doi.org/10.5301/JABFM.2012.9418] [PMID: 22865572]

[63]   Elisseeff J, Puleo C, Yang F, Sharma B. Advances in skeletal tissue engineering with hydrogels.
       Orthod Craniofac Res 2005; 8(3): 150-61.
       [http://dx.doi.org/10.1111/j.1601-6343.2005.00335.x] [PMID: 16022717]

[64]   Spiller KL, Maher SA, Lowman AM. Hydrogels for the repair of articular cartilage defects. Tissue
       Eng Part B Rev 2011; 17(4): 281-99.
       [http://dx.doi.org/10.1089/ten.teb.2011.0077] [PMID: 21510824]

[65]   Vinatier C, Guicheux J. Cartilage tissue engineering: From biomaterials and stem cells to osteoarthritis
       treatments. Ann Phys Rehabil Med 2016; S1877-0657(16)30001-X.

[66]   Place ES, George JH, Williams CK, Stevens MM. Synthetic polymer scaffolds for tissue engineering.
       Chem Soc Rev 2009; 38(4): 1139-51.
       [http://dx.doi.org/10.1039/b811392k] [PMID: 19421585]

[67]   Martin GR. Isolation of a pluripotent cell line from early mouse embryos cultured in medium
       conditioned by teratocarcinoma stem cells. Proc Natl Acad Sci USA 1981; 78(12): 7634-8.
       [http://dx.doi.org/10.1073/pnas.78.12.7634] [PMID: 6950406]

[68]   Toh WS, Lee EH, Cao T. Potential of human embryonic stem cells in cartilage tissue engineering and
       regenerative medicine. Stem Cell Rev 2011; 7(3): 544-59.
       [http://dx.doi.org/10.1007/s12015-010-9222-6] [PMID: 21188652]

[69]   Vats A, Bielby RC, Tolley N, *et al.* Chondrogenic differentiation of human embryonic stem cells: the
       effect of the micro-environment. Tissue Eng 2006; 12(6): 1687-97.
       [http://dx.doi.org/10.1089/ten.2006.12.1687] [PMID: 16846363]

[70]   zur Nieden NI, Kempka G, Rancourt DE, Ahr HJ. Induction of chondro-, osteo- and adipogenesis in
       embryonic stem cells by bone morphogenetic protein-2: effect of cofactors on differentiating lineages.
       BMC Dev Biol 2005; 5: 1.
       [http://dx.doi.org/10.1186/1471-213X-5-1] [PMID: 15673475]

[71]   Levenberg S, Huang NF, Lavik E, Rogers AB, Itskovitz-Eldor J, Langer R. Differentiation of human
       embryonic stem cells on three-dimensional polymer scaffolds. Proc Natl Acad Sci USA 2003;
       100(22): 12741-6.
       [http://dx.doi.org/10.1073/pnas.1735463100] [PMID: 14561891]

[72]   Hwang NS, Varghese S, Elisseeff J. Derivation of chondrogenically-committed cells from human
       embryonic cells for cartilage tissue regeneration. PLoS One 2008; 3(6): e2498.
       [http://dx.doi.org/10.1371/journal.pone.0002498] [PMID: 18575581]

[73]   Toh WS, Lee EH, Guo XM, *et al.* Cartilage repair using hyaluronan hydrogel-encapsulated human
       embryonic stem cell-derived chondrogenic cells. Biomaterials 2010; 31(27): 6968-80.
       [http://dx.doi.org/10.1016/j.biomaterials.2010.05.064] [PMID: 20619789]

[74]   Diekman BO, Christoforou N, Willard VP, Sun HS, Sanchez-Adams J, Leong KW. Cartilage tissue
       engineering using differentiated and purified induced pluripotent stem cells
       [http://dx.doi.org/10.1073/pnas.1210422109]

[75]   Brunger JM, Zutshi A, Willard VP, Gersbach CA, Guilak F. CRISPR/Cas9 editing of induced
       pluripotent stem cells for engineering inflammation-resistant tissues. Arthritis Rheumatol 2016; •••
       [http://dx.doi.org/10.1002/art.39982] [PMID: 27813286]

[76]   Diekman BO, Thakore PI, O'Connor SK, *et al.* Knockdown of the cell cycle inhibitor p21 enhances cartilage formation by induced pluripotent stem cells. Tissue Eng Part A 2015; 21(7-8): 1261-74.
[http://dx.doi.org/10.1089/ten.tea.2014.0240] [PMID: 25517798]

[77]   Türker T, Hassan K, Capdarest-Arest N. Extensor tendon gap reconstruction: a review. J Plast Surg Hand Surg 2016; 50(1): 1-6.
[http://dx.doi.org/10.3109/2000656X.2015.1086363] [PMID: 26400762]

[78]   Wong JK, Peck F. Improving results of flexor tendon repair and rehabilitation. Plast Reconstr Surg 2014; 134(6): 913e-25e.
[http://dx.doi.org/10.1097/PRS.0000000000000749] [PMID: 25415114]

[79]   Cukelj F, Bandalovic A, Knezevic J, Pavic A, Pivalica B, Bakota B. Treatment of ruptured Achilles tendon: Operative or non-operative procedure? Injury 2015; 46 (Suppl. 6): S137-42.
[http://dx.doi.org/10.1016/j.injury.2015.10.070] [PMID: 26573897]

[80]   Buller LT, Best MJ, Baraga MG, Kaplan LD. Trends in Anterior Cruciate Ligament Reconstruction in the United States. Orthop J Sports Med 2014; 3(1): 2325967114563664.
[PMID: 26535368]

[81]   Font Tellado S, Balmayor ER, Van Griensven M. Strategies to engineer tendon/ligament-to-bone interface: Biomaterials, cells and growth factors. Adv Drug Deliv Rev 2015; 94: 126-40.
[http://dx.doi.org/10.1016/j.addr.2015.03.004] [PMID: 25777059]

[82]   Chen X, Yin Z, Chen JL, *et al.* Scleraxis-overexpressed human embryonic stem cell-derived mesenchymal stem cells for tendon tissue engineering with knitted silk-collagen scaffold. Tissue Eng Part A 2014; 20(11-12): 1583-92.
[http://dx.doi.org/10.1089/ten.tea.2012.0656] [PMID: 24328506]

[83]   Orr SB, Chainani A, Hippensteel KJ, *et al.* Aligned multilayered electrospun scaffolds for rotator cuff tendon tissue engineering. Acta Biomater 2015; 24: 117-26.
[http://dx.doi.org/10.1016/j.actbio.2015.06.010] [PMID: 26079676]

[84]   Wong R, Alam N, McGrouther AD, Wong JK. Tendon grafts: their natural history, biology and future development. J Hand Surg Eur Vol 2015; 40(7): 669-81.
[http://dx.doi.org/10.1177/1753193415595176] [PMID: 26264585]

[85]   Zhang C, Yuan H, Liu H, *et al.* Well-aligned chitosan-based ultrafine fibers committed teno-lineage differentiation of human induced pluripotent stem cells for Achilles tendon regeneration. Biomaterials 2015; 53: 716-30.
[http://dx.doi.org/10.1016/j.biomaterials.2015.02.051] [PMID: 25890767]

[86]   Abbah SA, Spanoudes K, O'Brien T, Pandit A, Zeugolis DI. Assessment of stem cell carriers for tendon tissue engineering in pre-clinical models. Stem Cell Res Ther 2014; 5(2): 38.
[http://dx.doi.org/10.1186/scrt426] [PMID: 25157898]

[87]   Rui YF, Lui PP, Li G, Fu SC, Lee YW, Chan KM. Isolation and characterization of multipotent rat tendon-derived stem cells. Tissue Eng Part A 2010; 16(5): 1549-58.
[http://dx.doi.org/10.1089/ten.tea.2009.0529] [PMID: 20001227]

[88]   Zhang J, Wang JH. Characterization of differential properties of rabbit tendon stem cells and tenocytes. BMC Musculoskelet Disord 2010; 11: 10.
[http://dx.doi.org/10.1186/1471-2474-11-10] [PMID: 20082706]

[89]   Kristjánsson B, Limthongkul W, Yingsakmongkol W, Thantiworasit P, Jirathanathornnukul N, Honsawek S. Isolation and Characterization of Human Mesenchymal Stem Cells From Facet Joints and Interspinous Ligaments. Spine 2016; 41(1): E1-7.
[http://dx.doi.org/10.1097/BRS.0000000000001178] [PMID: 26555840]

[90]   Omae H, Zhao C, Sun YL, An KN, Amadio PC. Multilayer tendon slices seeded with bone marrow stromal cells: a novel composite for tendon engineering. J Orthop Res 2009; 27(7): 937-42.
[http://dx.doi.org/10.1002/jor.20823] [PMID: 19105224]

[91]   Zhang J, Li B, Wang JH. The role of engineered tendon matrix in the stemness of tendon stem cells *in vitro* and the promotion of tendon-like tissue formation *in vivo*. Biomaterials 2011; 32(29): 6972-81.
[http://dx.doi.org/10.1016/j.biomaterials.2011.05.088] [PMID: 21703682]

[92]   Yin Z, Chen X, Zhu T, *et al*. The effect of decellularized matrices on human tendon stem/progenitor cell differentiation and tendon repair. Acta Biomater 2013; 9(12): 9317-29.
[http://dx.doi.org/10.1016/j.actbio.2013.07.022] [PMID: 23896565]

[93]   Yao D, Liu H, Fan Y. Silk scaffolds for musculoskeletal tissue engineering. Exp Biol Med (Maywood) 2016; 241(3): 238-45.
[http://dx.doi.org/10.1177/1535370215606994] [PMID: 26445979]

[94]   Altman GH, Horan RL, Lu HH, *et al*. Silk matrix for tissue engineered anterior cruciate ligaments. Biomaterials 2002; 23(20): 4131-41.
[http://dx.doi.org/10.1016/S0142-9612(02)00156-4] [PMID: 12182315]

[95]   Hennecke K, Redeker J, Kuhbier JW, *et al*. Bundles of spider silk, braided into sutures, resist basic cyclic tests: potential use for flexor tendon repair. PLoS One 2013; 8(4): e61100.
[http://dx.doi.org/10.1371/journal.pone.0061100] [PMID: 23613793]

[96]   Musson DS, Naot D, Chhana A, *et al*. *In vitro* evaluation of a novel non-mulberry silk scaffold for use in tendon regeneration. Tissue Eng Part A 2015; 21(9-10): 1539-51.
[http://dx.doi.org/10.1089/ten.tea.2014.0128] [PMID: 25604072]

[97]   Chen K, Sahoo S, He P, Ng KS, Toh SL, Goh JC. A hybrid silk/RADA-based fibrous scaffold with triple hierarchy for ligament regeneration. Tissue Eng Part A 2012; 18(13-14): 1399-409.
[http://dx.doi.org/10.1089/ten.tea.2011.0376] [PMID: 22429111]

[98]   Hennecke K, Redeker J, Kuhbier JW, *et al*. Bundles of spider silk, braided into sutures, resist basic cyclic tests: potential use for flexor tendon repair. PLoS One 2013; 8(4): e61100.
[http://dx.doi.org/10.1371/journal.pone.0061100] [PMID: 23613793]

[99]   Tokareva O, Michalczechen-Lacerda VA, Rech EL, Kaplan DL. Recombinant DNA production of spider silk proteins. Microb Biotechnol 2013; 6(6): 651-63.
[http://dx.doi.org/10.1111/1751-7915.12081] [PMID: 24119078]

[100]   Shen W, Chen X, Hu Y, *et al*. Long-term effects of knitted silk-collagen sponge scaffold on anterior cruciate ligament reconstruction and osteoarthritis prevention. Biomaterials 2014; 35(28): 8154-63.
[http://dx.doi.org/10.1016/j.biomaterials.2014.06.019] [PMID: 24974007]

[101]   Moshaverinia A, Xu X, Chen C, *et al*. Application of stem cells derived from the periodontal ligament or gingival tissue sources for tendon tissue regeneration. Biomaterials 2014; 35(9): 2642-50.
[http://dx.doi.org/10.1016/j.biomaterials.2013.12.053] [PMID: 24397989]

[102]   Wagner ER, Bravo D, Dadsetan M, *et al*. Ligament Tissue Engineering Using a Novel Porous Polycaprolactone Fumarate Scaffold and Adipose Tissue-Derived Mesenchymal Stem Cells Grown in Platelet Lysate. Tissue Eng Part A 2015; 21(21-22): 2703-13.
[http://dx.doi.org/10.1089/ten.tea.2015.0183] [PMID: 26413793]

[103]   Lipner J, Shen H, Cavinatto L, *et al*. *In Vivo* Evaluation of Adipose-Derived Stromal Cells Delivered with a Nanofiber Scaffold for Tendon-to-Bone Repair. Tissue Eng Part A 2015; 21(21-22): 2766-74.
[http://dx.doi.org/10.1089/ten.tea.2015.0101] [PMID: 26414599]

[104]   Deng D, Wang W, Wang B, *et al*. Repair of Achilles tendon defect with autologous ASCs engineered tendon in a rabbit model. Biomaterials 2014; 35(31): 8801-9.
[http://dx.doi.org/10.1016/j.biomaterials.2014.06.058] [PMID: 25069604]

[105]   Zhang W, Yang Y, Zhang K, Li Y, Fang G. Weft-knitted silk-poly(lactide-co-glycolide) mesh scaffold combined with collagen matrix and seeded with mesenchymal stem cells for rabbit Achilles tendon repair. Connect Tissue Res 2015; 56(1): 25-34.
[http://dx.doi.org/10.3109/03008207.2014.976309] [PMID: 25333819]

[106] Sahoo S, Toh SL, Goh JC. A bFGF-releasing silk/PLGA-based biohybrid scaffold for ligament/tendon tissue engineering using mesenchymal progenitor cells. Biomaterials 2010; 31(11): 2990-8.
[http://dx.doi.org/10.1016/j.biomaterials.2010.01.004] [PMID: 20089300]

[107] Zhang J, Wang JH. Mechanobiological response of tendon stem cells: implications of tendon homeostasis and pathogenesis of tendinopathy. J Orthop Res 2010; 28(5): 639-43.
[PMID: 19918904]

[108] Xu Y, Dong S, Zhou Q, *et al.* The effect of mechanical stimulation on the maturation of TDSCs-poly(L-lactide-co-e-caprolactone)/collagen scaffold constructs for tendon tissue engineering. Biomaterials 2014; 35(9): 2760-72.
[http://dx.doi.org/10.1016/j.biomaterials.2013.12.042] [PMID: 24411676]

[109] Chen L, Dong SW, Liu JP, Tao X, Tang KL, Xu JZ. Synergy of tendon stem cells and platelet-rich plasma in tendon healing. J Orthop Res 2012; 30(6): 991-7.
[http://dx.doi.org/10.1002/jor.22033] [PMID: 22161871]

[110] Nakajima D, Tabata Y, Sato S. Periodontal tissue regeneration with PRP incorporated gelatin hydrogel sponges. Biomed Mater 2015; 10(5): 055016.
[http://dx.doi.org/10.1088/1748-6041/10/5/055016] [PMID: 26481592]

[111] Sánchez M, Anitua E, Azofra J, Andía I, Padilla S, Mujika I. Comparison of surgically repaired Achilles tendon tears using platelet-rich fibrin matrices. Am J Sports Med 2007; 35(2): 245-51.
[http://dx.doi.org/10.1177/0363546506294078] [PMID: 17099241]

[112] Giusti I, D'Ascenzo S, Manco A, Di Stefano G, Di Francesco M, Rughetti A, *et al.* Platelet concentration in platelet-rich plasma affects tenocyte behavior *in vitro*. Biomed Res Int 2014; 2014: 630870.

[113] Pan L, Yong Z, Yuk KS, Hoon KY, Yuedong S, Xu J. Growth Factor Release from Lyophilized Porcine Platelet-Rich Plasma: Quantitative Analysis and Implications for Clinical Applications. Aesthetic Plast Surg 2016; 40(1): 157-63.
[http://dx.doi.org/10.1007/s00266-015-0580-y] [PMID: 26516079]

[114] Garg K, Boppart MD. Influence of exercise and aging on extracellular matrix composition in the skeletal muscle stem cell niche. J Appl Physiol 2016; 121(5): 1053-8.
[http://dx.doi.org/10.1152/japplphysiol.00594.2016] [PMID: 27539500]

[115] Yin H, Price F, Rudnicki MA. Satellite cells and the muscle stem cell niche. Physiol Rev 2013; 93(1): 23-67.
[http://dx.doi.org/10.1152/physrev.00043.2011] [PMID: 23303905]

[116] Bosnakovski D, Xu Z, Li W, *et al.* Prospective isolation of skeletal muscle stem cells with a Pax7 reporter. Stem Cells 2008; 26(12): 3194-204.
[http://dx.doi.org/10.1634/stemcells.2007-1017] [PMID: 18802040]

[117] Mizuno H. The potential for treatment of skeletal muscle disorders with adipose-derived stem cells. Curr Stem Cell Res Ther 2010; 5(2): 133-6.
[http://dx.doi.org/10.2174/157488810791268573] [PMID: 19941455]

[118] Sarig R, Baruchi Z, Fuchs O, Nudel U, Yaffe D. Regeneration and transdifferentiation potential of muscle-derived stem cells propagated as myospheres. Stem Cells 2006; 24(7): 1769-78.
[http://dx.doi.org/10.1634/stemcells.2005-0547] [PMID: 16574751]

[119] Torrente Y, Belicchi M, Marchesi C, *et al.* Autologous transplantation of muscle-derived CD133+ stem cells in Duchenne muscle patients. Cell Transplant 2007; 16(6): 563-77.
[http://dx.doi.org/10.3727/000000007783465064] [PMID: 17912948]

[120] Dellavalle A, Sampaolesi M, Tonlorenzi R, *et al.* Pericytes of human skeletal muscle are myogenic precursors distinct from satellite cells. Nat Cell Biol 2007; 9(3): 255-67.
[http://dx.doi.org/10.1038/ncb1542] [PMID: 17293855]

[121] Darabi R, Gehlbach K, Bachoo RM, *et al.* Functional skeletal muscle regeneration from differentiating embryonic stem cells. Nat Med 2008; 14(2): 134-43.
[http://dx.doi.org/10.1038/nm1705] [PMID: 18204461]

[122] Darabi R, Santos FN, Filareto A, *et al.* Assessment of the myogenic stem cell compartment following transplantation of Pax3/Pax7-induced embryonic stem cell-derived progenitors. Stem Cells 2011; 29(5): 777-90.
[http://dx.doi.org/10.1002/stem.625] [PMID: 21374762]

[123] Darabi R, Arpke RW, Irion S, *et al.* Human ES- and iPS-derived myogenic progenitors restore DYSTROPHIN and improve contractility upon transplantation in dystrophic mice. Cell Stem Cell 2012; 10(5): 610-9.
[http://dx.doi.org/10.1016/j.stem.2012.02.015] [PMID: 22560081]

[124] Wilschut KJ, Haagsman HP, Roelen BA. Extracellular matrix components direct porcine muscle stem cell behavior. Exp Cell Res 2010; 316(3): 341-52.
[http://dx.doi.org/10.1016/j.yexcr.2009.10.014] [PMID: 19853598]

[125] Porzionato A, Sfriso MM, Pontini A, *et al.* Decellularized human skeletal muscle as biologic scaffold for reconstructive surgery. Int J Mol Sci 2015; 16(7): 14808-31.
[http://dx.doi.org/10.3390/ijms160714808] [PMID: 26140375]

[126] Choi YJ, Kim TG, Jeong J, *et al.* 3D cell printing of functional skeletal muscle constructs using skeletal muscle-derived bioink. Adv Healthc Mater 2016; 5(20): 2636-45.
[http://dx.doi.org/10.1002/adhm.201600483] [PMID: 27529631]

[127] Chevallay B, Herbage D. Collagen-based biomaterials as 3D scaffold for cell cultures: applications for tissue engineering and gene therapy. Med Biol Eng Comput 2000; 38(2): 211-8.
[http://dx.doi.org/10.1007/BF02344779] [PMID: 10829416]

[128] Chattopadhyay S, Raines RT. Review Quantifying the relation between bond number and myoblast proliferation wound healing. Biopolymers 2014; 101(8): 821-33.
[http://dx.doi.org/10.1002/bip.22486] [PMID: 24633807]

[129] Thorrez L, Shansky J, Wang L, *et al.* Growth, differentiation, transplantation and survival of human skeletal myofibers on biodegradable scaffolds. Biomaterials 2008; 29(1): 75-84.
[http://dx.doi.org/10.1016/j.biomaterials.2007.09.014] [PMID: 17928049]

[130] Carnio S, Serena E, Rossi CA, De Coppi P, Elvassore N, Vitiello L. Three-dimensional porous scaffold allows long-term wild-type cell delivery in dystrophic muscle. J Tissue Eng Regen Med 2011; 5(1): 1-10.
[http://dx.doi.org/10.1002/term.282] [PMID: 20607681]

[131] Serena E, Flaibani M, Carnio S, *et al.* Electrophysiologic stimulation improves myogenic potential of muscle precursor cells grown in a 3D collagen scaffold. Neurol Res 2008; 30(2): 207-14.
[http://dx.doi.org/10.1179/174313208X281109] [PMID: 18397614]

[132] Kroehne V, Heschel I, Schügner F, Lasrich D, Bartsch JW, Jockusch H. Use of a novel collagen matrix with oriented pore structure for muscle cell differentiation in cell culture and in grafts. J Cell Mol Med 2008; 12(5A): 1640-8.
[http://dx.doi.org/10.1111/j.1582-4934.2008.00238.x] [PMID: 18194451]

[133] van Wachem PB, Plantinga JA, Wissink MJ, *et al.* *In vivo* biocompatibility of carbodiimide-crosslinked collagen matrices: Effects of crosslink density, heparin immobilization, and bFGF loading. J Biomed Mater Res 2001; 55(3): 368-78.
[http://dx.doi.org/10.1002/1097-4636(20010605)55:3<368::AID-JBM1025>3.0.CO;2-5] [PMID: 11255190]

[134] Makridakis JL, Pins GD, Dominko T, Page RL. Design of a novel engineered muscle construct using muscle derived fibroblastic cells seeded onto braided collagen threads, in: Bioengineering Conference, 2009 IEEE 35th Annual Northeast 2009; 1-2.

[http://dx.doi.org/10.1109/NEBC.2009.4967673]

[135] Frey SP, Jansen H, Raschke MJ, Meffert RH, Ochman S. VEGF improves skeletal muscle regeneration after acute trauma and reconstruction of the limb in a rabbit model. Clin Orthop Relat Res 2012; 470(12): 3607-14.
[http://dx.doi.org/10.1007/s11999-012-2456-7] [PMID: 22806260]

[136] Ju YM, Atala A, Yoo JJ, Lee SJ. *In situ* regeneration of skeletal muscle tissue through host cell recruitment. Acta Biomater 2014; 10(10): 4332-9.
[http://dx.doi.org/10.1016/j.actbio.2014.06.022] [PMID: 24954910]

[137] Vandenburgh H, Shansky J, Benesch-Lee F, *et al.* Drug-screening platform based on the contractility of tissue-engineered muscle. Muscle Nerve 2008; 37(4): 438-47.
[http://dx.doi.org/10.1002/mus.20931] [PMID: 18236465]

[138] Rowley JA, Madlambayan G, Mooney DJ. Alginate hydrogels as synthetic extracellular matrix materials. Biomaterials 1999; 20(1): 45-53.
[http://dx.doi.org/10.1016/S0142-9612(98)00107-0] [PMID: 9916770]

[139] Boontheekul T, Kong HJ, Mooney DJ. Controlling alginate gel degradation utilizing partial oxidation and bimodal molecular weight distribution. Biomaterials 2005; 26(15): 2455-65.
[http://dx.doi.org/10.1016/j.biomaterials.2004.06.044] [PMID: 15585248]

[140] Drury JL, Boontheekul T, Mooney DJ. Cellular cross-linking of peptide modified hydrogels. J Biomech Eng 2005; 127(2): 220-8.
[http://dx.doi.org/10.1115/1.1865194] [PMID: 15971699]

[141] Wang L, Shansky J, Borselli C, Mooney D, Vandenburgh H. Design and fabrication of a biodegradable, covalently crosslinked shape-memory alginate scaffold for cell and growth factor delivery. Tissue Eng Part A 2012; 18(19-20): 2000-7.
[http://dx.doi.org/10.1089/ten.tea.2011.0663] [PMID: 22646518]

[142] Boontheekul T, Hill EE, Kong HJ, Mooney DJ. Regulating myoblast phenotype through controlled gel stiffness and degradation. Tissue Eng 2007; 13(7): 1431-42.
[http://dx.doi.org/10.1089/ten.2006.0356] [PMID: 17561804]

[143] Rowley JA, Mooney DJ. Alginate type and RGD density control myoblast phenotype. J Biomed Mater Res 2002; 60(2): 217-23.
[http://dx.doi.org/10.1002/jbm.1287] [PMID: 11857427]

[144] Boontheekul T, Kong HJ, Hsiong SX, *et al.* Quantifying the relation between bond number and myoblast proliferation. Faraday Discuss 2008; 139: 53-70.
[http://dx.doi.org/10.1039/b719928g]

[145] Hill E, Boontheekul T, Mooney DJ. Designing scaffolds to enhance transplanted myoblast survival and migration. Tissue Eng 2006; 12(5): 1295-304.
[http://dx.doi.org/10.1089/ten.2006.12.1295] [PMID: 16771642]

[146] Hill E, Boontheekul T, Mooney DJ. Regulating activation of transplanted cells controls tissue regeneration. Proc Natl Acad Sci USA 2006; 103(8): 2494-9.
[http://dx.doi.org/10.1073/pnas.0506004103] [PMID: 16477029]

[147] Borselli C, Storrie H, Benesch-Lee F, *et al.* Functional muscle regeneration with combined delivery of angiogenesis and myogenesis factors. Proc Natl Acad Sci USA 2010; 107(8): 3287-92.
[http://dx.doi.org/10.1073/pnas.0903875106] [PMID: 19966309]

[148] Jana S, Cooper A, Zhang M. Chitosan scaffolds with unidirectional microtubular pores for large skeletal myotube generation. Adv Healthc Mater 2013; 2(4): 557-61.
[http://dx.doi.org/10.1002/adhm.201200177] [PMID: 23184507]

[149] Rossi CA, Flaibani M, Blaauw B, *et al. In vivo* tissue engineering of functional skeletal muscle by freshly isolated satellite cells embedded in a photopolymerizable hydrogel. FASEB J 2011; 25(7): 2296-304.

[http://dx.doi.org/10.1096/fj.10-174755] [PMID: 21450908]

[150] Wang LS, Boulaire J, Chan PP, Chung JE, Kurisawa M. The role of stiffness of gelatin-hydroxyphenylpropionic acid hydrogels formed by enzyme-mediated crosslinking on the differentiation of human mesenchymal stem cell. Biomaterials 2010; 31(33): 8608-16.
[http://dx.doi.org/10.1016/j.biomaterials.2010.07.075] [PMID: 20709390]

[151] Nelson DM, Baraniak PR, Ma Z, Guan J, Mason NS, Wagner WR. Controlled release of IGF-1 and HGF from a biodegradable polyurethane scaffold. Pharm Res 2011; 28(6): 1282-93.
[http://dx.doi.org/10.1007/s11095-011-0391-z] [PMID: 21347565]

[152] Hoque ME, San WY, Wei F, *et al.* Processing of polycaprolactone and polycaprolactone-based copolymers into 3D scaffolds, and their cellular responses. Tissue Eng Part A 2009; 15(10): 3013-24.
[http://dx.doi.org/10.1089/ten.tea.2008.0355] [PMID: 19331580]

[153] Guelcher SA. Biodegradable polyurethanes: synthesis and applications in regenerative medicine. Tissue Eng Part B Rev 2008; 14(1): 3-17.
[http://dx.doi.org/10.1089/teb.2007.0133] [PMID: 18454631]

[154] Hurd SA, Bhatti NM, Walker AM, Kasukonis BM, Wolchok JC. Development of a biological scaffold engineered using the extracellular matrix secreted by skeletal muscle cells. Biomaterials 2015; 49: 9-17.
[http://dx.doi.org/10.1016/j.biomaterials.2015.01.027] [PMID: 25725550]

[155] Cronin EM, Thurmond FA, Bassel-Duby R, *et al.* Protein-coated poly(L-lactic acid) fibers provide a substrate for differentiation of human skeletal muscle cells. J Biomed Mater Res A 2004; 69(3): 373-81.
[http://dx.doi.org/10.1002/jbm.a.30009] [PMID: 15127383]

[156] Saxena AK, Marler J, Benvenuto M, Willital GH, Vacanti JP. Skeletal muscle tissue engineering using isolated myoblasts on synthetic biodegradable polymers: preliminary studies. Tissue Eng 1999; 5(6): 525-32.
[http://dx.doi.org/10.1089/ten.1999.5.525] [PMID: 10611544]

[157] Saxena AK, Willital GH, Vacanti JP. Vascularized three-dimensional skeletal muscle tissue-engineering. Biomed Mater Eng 2001; 11(4): 275-81.
[PMID: 11790859]

[158] Bandyopadhyay B, Shah V, Soram M, Viswanathan C, Ghosh D. *In vitro* and *in vivo* evaluation of (L)-lactide/ε-caprolactone copolymer scaffold to support myoblast growth and differentiation. Biotechnol Prog 2013; 29(1): 197-205.
[http://dx.doi.org/10.1002/btpr.1665] [PMID: 23143919]

[159] Peters MC, Polverini PJ, Mooney DJ. Engineering vascular networks in porous polymer matrices. J Biomed Mater Res 2002; 60(4): 668-78.
[http://dx.doi.org/10.1002/jbm.10134] [PMID: 11948526]

[160] Harris LD, Kim BS, Mooney DJ. Open pore biodegradable matrices formed with gas foaming. J Biomed Mater Res 1998; 42(3): 396-402.
[http://dx.doi.org/10.1002/(SICI)1097-4636(19981205)42:3<396::AID-JBM7>3.0.CO;2-E] [PMID: 9788501]

[161] Smith MK, Peters MC, Richardson TP, Garbern JC, Mooney DJ. Locally enhanced angiogenesis promotes transplanted cell survival. Tissue Eng 2004; 10(1-2): 63-71.
[http://dx.doi.org/10.1089/107632704322791709] [PMID: 15009931]

[162] Yu J, Takanari K, Hong Y, *et al.* Non-invasive characterization of polyurethane-based tissue constructs in a rat abdominal repair model using high frequency ultrasound elasticity imaging. Biomaterials 2013; 34(11): 2701-9.
[http://dx.doi.org/10.1016/j.biomaterials.2013.01.036] [PMID: 23347836]

[163] Choi JS, Lee SJ, Christ GJ, Atala A, Yoo JJ. The influence of electrospun aligned poly(epsilon-

caprolactone)/collagen nanofiber meshes on the formation of self-aligned skeletal muscle myotubes. Biomaterials 2008; 29(19): 2899-906.
[http://dx.doi.org/10.1016/j.biomaterials.2008.03.031] [PMID: 18400295]

[164] Guex AG, Birrer DL, Fortunato G, Tevaearai HT, Giraud MN. Anisotropically oriented electrospun matrices with an imprinted periodic micropattern: a new scaffold for engineered muscle constructs. Biomed Mater 2013; 8(2): 021001.
[http://dx.doi.org/10.1088/1748-6041/8/2/021001] [PMID: 23343525]

[165] Ku SH, Lee SH, Park CB. Synergic effects of nanofiber alignment and electroactivity on myoblast differentiation. Biomaterials 2012; 33(26): 6098-104.
[http://dx.doi.org/10.1016/j.biomaterials.2012.05.018] [PMID: 22681977]

[166] Chen MC, Sun YC, Chen YH. Electrically conductive nanofibers with highly oriented structures and their potential application in skeletal muscle tissue engineering. Acta Biomater 2013; 9(3): 5562-72.
[http://dx.doi.org/10.1016/j.actbio.2012.10.024] [PMID: 23099301]

[167] Yang J, Jao B, McNally AK, Anderson JM. *In vivo* quantitative and qualitative assessment of foreign body giant cell formation on biomaterials in mice deficient in natural killer lymphocyte subsets, mast cells, or the interleukin-4 receptorα and in severe combined immunodeficient mice. J Biomed Mater Res A 2014; 102(6): 2017-23.
[http://dx.doi.org/10.1002/jbm.a.35152] [PMID: 24616384]

[168] McNally AK, Anderson JM. Phenotypic expression in human monocyte-derived interleukin-4-induced foreign body giant cells and macrophages *in vitro*: dependence on material surface properties. J Biomed Mater Res A 2015; 103(4): 1380-90.
[http://dx.doi.org/10.1002/jbm.a.35280] [PMID: 25045023]

[169] Anderson JM, Rodriguez A, Chang DT. Foreign body reaction to biomaterials. Semin Immunol 2008; 20(2): 86-100.
[http://dx.doi.org/10.1016/j.smim.2007.11.004] [PMID: 18162407]

[170] Klinge U, Klosterhalfen B, Müller M, Schumpelick V. Foreign body reaction to meshes used for the repair of abdominal wall hernias. Eur J Surg 1999; 165(7): 665-73.
[http://dx.doi.org/10.1080/11024159950189726] [PMID: 10452261]

[171] Klopfleisch R, Jung F. The pathology of the foreign body reaction against biomaterials. J Biomed Mater Res A 2017; 105(3): 927-40.
[http://dx.doi.org/10.1002/jbm.a.35958] [PMID: 27813288]

[172] Bosworth LA, Turner LA, Cartmell SH. State of the art composites comprising electrospun fibres coupled with hydrogels: a review. Nanomedicine (Lond) 2013; 9(3): 322-35.
[http://dx.doi.org/10.1016/j.nano.2012.10.008] [PMID: 23178282]

[173] Sahoo S, DeLozier KR, Dumm RA, Rosen MJ, Derwin KA. Fiber-reinforced dermis graft for ventral hernia repair. J Mech Behav Biomed Mater 2014; 34: 320-9.
[http://dx.doi.org/10.1016/j.jmbbm.2014.03.001] [PMID: 24704969]

[174] Sirivisoot S, Pareta R, Harrison BS. Protocol and cell responses in three-dimensional conductive collagen gel scaffolds with conductive polymer nanofibres for tissue regeneration. Interface Focus 2014; 4(1): 20130050.
[http://dx.doi.org/10.1098/rsfs.2013.0050] [PMID: 24501678]

[175] Jana S, Leung M, Chang J, Zhang M. Effect of nano- and micro-scale topological features on alignment of muscle cells and commitment of myogenic differentiation. Biofabrication 2014; 6(3): 035012.
[http://dx.doi.org/10.1088/1758-5082/6/3/035012] [PMID: 24876344]

[176] Leung M, Cooper A, Jana S, Tsao CT, Petrie TA, Zhang M. Nanofiber-based *in vitro* system for high myogenic differentiation of human embryonic stem cells. Biomacromolecules 2013; 14(12): 4207-16.
[http://dx.doi.org/10.1021/bm4009843] [PMID: 24131307]

[177]   Hong Y, Takanari K, Amoroso NJ, *et al.* An elastomeric patch electrospun from a blended solution of dermal extracellular matrix and biodegradable polyurethane for rat abdominal wall repair. Tissue Eng Part C Methods 2012; 18(2): 122-32.
[http://dx.doi.org/10.1089/ten.tec.2011.0295] [PMID: 21933017]

[178]   Hong Y, Huber A, Takanari K, *et al.* Mechanical properties and *in vivo* behavior of a biodegradable synthetic polymer microfiber-extracellular matrix hydrogel biohybrid scaffold. Biomaterials 2011; 32(13): 3387-94.
[http://dx.doi.org/10.1016/j.biomaterials.2011.01.025] [PMID: 21303718]

[179]   van 't Riet M, de Vos van Steenwijk PJ, Bonthuis F, *et al.* Prevention of adhesion to prosthetic mesh: comparison of different barriers using an incisional hernia model. Ann Surg 2003; 237(1): 123-8.
[http://dx.doi.org/10.1097/00000658-200301000-00017] [PMID: 12496539]

[180]   Haslauer CM, Moghe AK, Osborne JA, Gupta BS, Loboa EG. Collagen-PCL sheath-core bicomponent electrospun scaffolds increase osteogenic differentiation and calcium accretion of human adipose-derived stem cells. J Biomater Sci Polym Ed 2011; 22(13): 1695-712.
[http://dx.doi.org/10.1163/092050610X521595] [PMID: 20836922]

[181]   Kim MS, Jun I, Shin YM, Jang W, Kim SI, Shin H. The development of genipin-crosslinked poly(caprolactone) (PCL)/gelatin nanofibers for tissue engineering applications. Macromol Biosci 2010; 10(1): 91-100.
[http://dx.doi.org/10.1002/mabi.200900168] [PMID: 19685497]

[182]   Shen Z, Guo S, Ye D, *et al.* Skeletal muscle regeneration on protein-grafted and microchannel-patterned scaffold for hypopharyngeal tissue engineering. BioMed Res Int 2013; 2013: 146953.
[http://dx.doi.org/10.1155/2013/146953] [PMID: 24175281]

[183]   Wolf MT, Carruthers CA, Dearth CL, *et al.* Polypropylene surgical mesh coated with extracellular matrix mitigates the host foreign body response. J Biomed Mater Res A 2014; 102(1): 234-46.
[http://dx.doi.org/10.1002/jbm.a.34671] [PMID: 23873846]

[184]   Lesman A, Koffler J, Atlas R, Blinder YJ, Kam Z, Levenberg S. Engineering vessel-like networks within multicellular fibrin-based constructs. Biomaterials 2011; 32(31): 7856-69.
[http://dx.doi.org/10.1016/j.biomaterials.2011.07.003] [PMID: 21816465]

[185]   Borrazzo EC, Belmont MF, Boffa D, Fowler DL. Effect of prosthetic material on adhesion formation after laparoscopic ventral hernia repair in a porcine model. Hernia 2004; 8(2): 108-12.
[http://dx.doi.org/10.1007/s10029-003-0181-6] [PMID: 14634842]

[186]   Schreinemacher MH, Emans PJ, Gijbels MJ, Greve JW, Beets GL, Bouvy ND. Degradation of mesh coatings and intraperitoneal adhesion formation in an experimental model. Br J Surg 2009; 96(3): 305-13.
[http://dx.doi.org/10.1002/bjs.6446] [PMID: 19224521]

[187]   Pierce RA, Perrone JM, Nimeri A, *et al.* 120-day comparative analysis of adhesion grade and quantity, mesh contraction, and tissue response to a novel omega-3 fatty acid bioabsorbable barrier macroporous mesh after intraperitoneal placement. Surg Innov 2009; 16(1): 46-54.
[http://dx.doi.org/10.1177/1553350608330479] [PMID: 19124448]

[188]   Zhou W, He DQ, Liu JY, *et al.* Angiogenic gene-modified myoblasts promote vascularization during repair of skeletal muscle defects. J Tissue Eng Regen Med 2015; 9(12): 1404-16.
[http://dx.doi.org/10.1002/term.1692] [PMID: 23365046]

[189]   Yin T, Wang G, He S, *et al.* Malignant Pleural Effusion and Ascites Induce Epithelial-Mesenchymal Transition and Cancer Stem-Like Cell Properties *via* VEGF/PI3K/Akt/mTOR Pathway. J Biol Chem 2016; pii: jbc.M116.753236.

[190]   Mirahmadi M, Ahmadiankia N, Naderi-Meshkin H, *et al.* Hypoxia and laser enhance expression of SDF-1 in muscles cells. Cell Mol Biol 2016; 62(5): 31-7.
[PMID: 27188867]

[191] Zordan P, Rigamonti E, Freudenberg K, *et al.* Macrophages commit postnatal endothelium-derived progenitors to angiogenesis and restrict endothelial to mesenchymal transition during muscle regeneration. Cell Death Dis 2014; 5: e1031.
[http://dx.doi.org/10.1038/cddis.2013.558] [PMID: 24481445]

[192] Juhas M, Engelmayr GC Jr, Fontanella AN, Palmer GM, Bursac N. Biomimetic engineered muscle with capacity for vascular integration and functional maturation *in vivo.* Proc Natl Acad Sci USA 2014; 111(15): 5508-13.
[http://dx.doi.org/10.1073/pnas.1402723111] [PMID: 24706792]

[193] Madden L, Juhas M, Kraus WE, Truskey GA, Bursac N. Bioengineered human myobundles mimic clinical responses of skeletal muscle to drugs. eLife 2015; 4: e04885.
[http://dx.doi.org/10.7554/eLife.04885] [PMID: 25575180]

SUBJECT INDEX

A

B

Mehdi Razavi (Ed.)
All rights reserved-© 2017 Bentham Science Publishers

www.ingramcontent.com/pod-product-compliance
Lightning Source LLC
Chambersburg PA
CBHW041727210326
41598CB00008B/807

* 9 7 8 1 6 8 1 0 8 5 7 9 1 *